TRAVELS WITH PETE

By Kris Fischer

Wilmington, DE

Copyright © 2025 Kris Fischer

All rights reserved. No portion of this book may be reproduced mechanically, electronically, or by any other means, including photocopying, without written permission from the author. It is illegal to copy this book, post it to a website, or distribute it by any other means without permission from the author.

Author Contact: KrisFischer.com

Thomas Noble Books
Wilmington, DE

ISBN: 978-1-945586-33-0

First Printing: 2025

For more information and to view photos of the journey described in this book, please visit: KrisFischer.com

This publication is designed to provide accurate and authoritative information regarding the subject matter covered. It is sold with the understanding that the author is not engaged in rendering professional services. If travel, legal, accounting, medical, psychological, or any other expert assistance is required, the services of a competent professional should be sought.

TABLE OF CONTENTS

Chapter 1: Under the Southern Sky 7

Chapter 2: Sleeping with the Miners 13

Chapter 3: Catch a Falling Star 19

Chapter 4: Stars Align 27

Chapter 5: Pete's Accident 35

Chapter 6: Wedding Down Under Dec. 15, 1973 43

Chapter 7: Crocodiles and Floods 53

Chapter 8: How Deep is Your Love? 57

Chapter 9: Thursday Island University: Beer, Women, Shrimp ... 63

Chapter 10: New Guinea's Colorful Characters 71

Chapter 11: Balinese Shangri-La 77

Chapter 12: Batikking in Jogjakarta 87

Chapter 13: Singapore Slings at Raffles
 and Ribs at the Petroleum Club 95

Chapter 14: Breathtaking Beaches, Jungle, Pirates,
 and Barbed Wire Barricades 101

Chapter 15: Sleeping with My Husband
 in a Penang Whorehouse 113

Chapter 16: Lake Toba, Sumatra's Forever Paradise 123

Chapter 17: Land of the "King and I" 131

Chapter 18: Cigars, Jade, and Orchids 141

Chapter 19: At the Top of the World 151

Chapter 20: Trekking the Kali Gandaki River Valley 157

Chapter 21: Lepers, a Living Goddess, and Rosie the Brit 169

Chapter 22: Washing Away Ten Lifetimes of Sins 175

Chapter 23: Taj Mahal: a Labor of Love 183

Chapter 24: From the Desert to the Metropolis 191

Chapter 25: Sitting Still for a While 203

Chapter 26: Venice of the East: Srinagar, Kashmir 209

Chapter 27: Holiest Site of Sikhism enroute to Lahore, Pakistan 215

Chapter 28: Welcome to Afghanistan! 221

Chapter 29: From the Eyes of the Buddha 229

Chapter 30: Racing across Afghanistan and Iran into Turkey 235

Chapter 31: Where East Meets West 245

Chapter 32: Going Behind the Iron Curtain 253

Chapter 33: Culture Shock 265

Chapter 34: Meeting up with Old Friends in
              Holland & Scotland 275

Chapter 35: Homecoming 283

Epilogue 289

About the Author 290

Acknowledgements 291

Hornell Evening Tribune

Hornell, NY
Hornell Girl Heads to Australia

Krissy Rohver, daughter of Louise and Paul Rohver of North Hornell flew from Los Angeles, California, to Sydney, Australia, Thursday, January 20 1972, on Qantas Flight 567. Miss Rohver graduated from Hornell Senior High School in 1967. In December She earned a BSc degree from University of Maryland.

Las Virgenes Press

Calabasas, CA
Calabasas Boy Leaves for Australia

Peter Fischer, son of Steen and Mary Anne Fischer of Stokes Canyon Road sailed for Sydney, Australia, Thursday, March 9 1972 aboard the Danish freighter, Cap Coville. Peter was a member of the first graduating class of Agoura High School in May, 1967. In January he graduated from the University of California at Santa Barbara.

Two small-town kids from opposite sides of the US head off to see the world with a few dollars and some clothes stuffed in backpacks. A year later they will meet in Tasmania, Australia.

To follow their journey in photos, visit KrisFischer.com

CHAPTER 1

Under the Southern Sky

The train jolted forward, and I knew this trip would change everything, but not in the way I imagined. Haley, diagnosed with walking pneumonia, had to stay behind with Rhonda, leaving me to travel Tasmania alone. I barely notice the sleeper car's Victorian décor, as my mind spins with both worry for Haley and excitement about what awaits me in Tasmania.

I'd never traveled alone before and it both scares and challenges me. Where do I go? How far is the train station from the airport? What should you pay for a taxi? What if I lose my luggage? These are problems I'll have to solve for myself. All I know about Tasmania is that it lies off the southernmost tip of Australia and resembles New England in its weather, ocean, and Victorian architecture.

Twelve months ago, I had arrived in Sydney with my college roommate, Haley. We are now teaching in the city suburbs of Sydney. The plan had been to take a week-long dash around Tasmania with Haley and our fellow teacher, Rhonda.

The carriage sways back and forth as I look out the window at my twenty-three-year-old self staring back at me. Strands of long brown hair fall over my wide, blue eyes and tanned face. I smile with anticipation of a new place and a new adventure.

Flying into Launceston from Melbourne, what I remember most vividly is how green Tasmania looks. After spending the year in Sydney's Mediterranean climate, with its brown and dry landscape, tough succulents and gum trees, it's wonderful to see soft-petaled flowers and deciduous trees. Two rivers cojoin in the city. Along their banks perch well-kept wooden houses and quaint shops. I window shop as I head to the local hostel.

Standing at the check-in desk at the hostel, the host leans across the counter and whispers, "There's another Yank staying here. You ought 'ta meet him."

"Right! I didn't travel fifteen thousand miles from the US to meet another Yank. I could have stayed home to do that!" Just then, a gaggle of young travelers burst through the door. I know right away which one is the Yank.

He's tall and lean with a Western cowboy build; broad shoulders, and narrow hips. He wears ill-fitting clothes with wide gaps of skin showing around his ankles and wrists. He needs a haircut. Dirty blond hair tufts out behind his ears. He's wearing thick black glasses like the ones Garry Moore, an early TV host, wore in the fifties. They dominate his face, but his straight-edged smile and blue eyes are welcoming.

He looks so young that I think at first, he's a high school dropout, but I find out later he's a graduate of the University of California at Santa Barbara, UCSB.

After dumping my backpack in the dorm, I join the others for dinner. The usual small talk follows. "Where are you from?" "What do you do?" "Where did you go to school?" "Where have you been in Tasmania?"

Although I initially resisted being paired up with the Yank, I end up mainly talking with him. His name is Pete. I find out he's from California.

"Hey, last summer, I went to Santa Barbara to make up some history credits. I've dreamed about California since I was seven, watching 'The Wonderful World of Disney' on Saturday nights with my family in Hornell. That sunshine always looked so good compared to all the snow and cold in western New York."

Pete replies, "I grew up in the San Fernando Valley and then moved to Calabasas with my six younger brothers and sisters. Agoura High School was as new as our house. California isn't so special when you live there full-time. The weather can be boring. I'd love to live through a winter in Montana."

"You know, Pete, I wouldn't have graduated if I'd gone to UCSB and not the University of Maryland," I add, "All that beach and ocean right there to play in all day. Isla Vista was a trip. I must have eaten the dollar dinners at every restaurant in town. I lived in an apartment right on the beach overlooking the ocean and two blocks from campus, Del Playa Drive."

"Wow, you lived on the same street as my friend, Niels Nyborg! Well, he lived in his van on that street," Pete says.

"Are you kidding? I know Niels! He parked in front of our apartment. He knew my roommate, Shannon. What a coincidence," I say. "Here we are half way around the world with a friend in common!"

The conversation flows easily all night. I find out Pete had stayed in Sydney with a brain surgeon friend of his dad's, Fred Street. Fred's son almost killed Pete by taking him scuba diving in Sydney Harbor with a half-full tank of air, which ran out.

From Sydney, Pete had gone to the Northern Territory and worked as a jack-a-roo (young Australian cowboy) at Brunette Downs. Then he trapped rabbits in New South Wales with a professional rabbit trapper, clerked in a Melbourne office, and made pallets to hold newsprint at a paper mill in Bernie, Tasmania. Now, as he puts it, he has the best-paying job of his life as a miner in Roseberry, Tasmania, at the Electrolytic Zinc Company of Australasia.

My first year in Australia seems pretty tame compared to Pete's. I emigrated to Australia from the US with Haley. The Australian government paid our way over. In exchange, we had to stay and teach for two years. Australia wanted teachers and women in 1972. We wanted adventure and jobs.

At that time, Australia had a whites-only immigration policy and a shortage of females. Aussie men popularly believed the lack of women was because the American GIs stole their women during WW II while stationed in Australia.

Australia was also revamping its education training system from a two-year to four-year-degree teacher requirement, so foreign teachers filled the gap while Aussie teachers got more training. The government was betting we'd fall in love with handsome Aussies, marry them, and settle in Australia.

So far, I'd driven to Alice Springs and Ayer's Rock over a long weekend, spent spring break on a Hayman Island resort in the Great Barrier Reef, and been to the Blue Mountains. These trips were tame tourist destinations and couldn't compare with Pete's adventures.

As we say good night, Pete volunteers, "You and your girlfriends ought to drive along the western coast of Tasmania on your way to Hobart. It's an unexplored area with virgin forests, strange animals, and a spectacular coastline. Rosebery is on the way. I could find a free place for all of you to sleep."

Free always sounds good when you're traveling on the cheap, but I have to admit, I'm also interested in Pete, the vagabond miner. Pete has charmed me with his boyish smile and storytelling. His humor and goodwill are infectious. I never suspect that he'll soon steal my heart.

CHAPTER 2

Sleeping with the Miners

I'm standing in front of the Launceston hostel waiting for my girlfriends when I see a car speeding down the street toward me. A snappy, red Morris Mini pulls up in front, and Haley and Rhonda hop out.

Rhonda loves stick-shift cars, so we're all probably paying more rental bucks for this sporty number. Rhonda fancies herself a race-car driver. She goes too fast, but since she's a native Australian, we let her drive.

Unfortunately, we Yanks have trouble navigating on the "wrong" side of the road and other Aussie traffic anomalies, like giving way to the right. A car can back out of a driveway onto a highway without looking, and you're in the wrong if you don't give way. So, if we drive, it's more dangerous than Rhonda's speeding.

Haley's rosy cheeks are bright against her freckled face, framed by a red kerchief and her black, curly hair. Rhonda is short and athletic with light-brown hair cut in a shag. Both are puffing on cigarettes. Rhonda yahoos. "'Ow are ya, mate?"

"Great," I say. "You're both looking fit."

Haley answers, "Amazing what antibiotics can do. I should have listened to you about going to the doc sooner, but I'm totally on the mend now. What's our plan? A night here in Launceston and then on to Hobart?"

"Park and get your things, and we'll talk about it over lunch," I say. They park the car, check into the hostel, and go to a nearby cafe.

"I thought we should take the scenic route to Hobart down the west coast. I met a Yank from California working in a mine in Rosebery which is on our way to Hobart. He's offered us a free place to sleep in Rosebery. He says the west coast of Tasmania is undiscovered and wild. We shouldn't miss it," I tell them.

Rhonda and Haley roll their eyes and smile.

"What?" I ask.

"What a surprise, Rohver, that you met a guy," Haley says. (It's always bad when Haley uses my last name) "You've been picking up guys all year long. First, you accepted a date with the Qantas steward before we even arrived in Australia. Then Tom Cottee. You were waitressing in the restaurant on top of Sydney's Circular Key while we were waiting for our teaching credentials to arrive when you picked up that stuck-up prig. Heir to the Cottee Food millions, he wore a frickin' ascot every time he took you out!"

Rhonda chimes in, "Remember the Canadian hockey player? He brought you that old heavy surfboard that you tried to use and almost had a concussion when it thwacked you over the head at Cronulla Beach."

"Then there was that other Canadian, Bryon," Haley adds. "He was your boss when you were his receptionist at his sleazy, pyramid-scheme business in Rose Bay."

"Next was Ken Denham, the teacher at your school," Rhonda says. "You broke his heart. He thought that dating meant you were engaged. Made things a bit tough at work, eh? Especially when you started dating another guy at the same time."

My face turns red as they talk about my dating. "I guess you're right. I've been acting crazy this year. Pete, the Yank, is truly just another traveler. No dating. No romance. I'm swearing off men for my New Year's resolution," I answer.

Haley retorts, "Sure, Rohver. Didn't you just accept a New Year's Eve invite from Ian, the rugby-playing engineer from New South Wales University? The guy who you made out with under the table at the university toga party? That sounds like you're still in the dating game to me."

Haley takes another drag off her cigarette and continues. "Just a year ago, you were deciding whether to marry Ron Fountain, the only guy you dated in college, or come to Australia with me. It's a good thing you came to Australia. You obviously needed to date some more."

"Thanks for keeping such good track of my social life. You'd have more of a social life if you weren't so picky. I know it's hard to believe from my recent history, but Pete's just a friend. He made me laugh. Was he good-looking? I can't tell you because I can't remember. As to the guy counts, I'm pretty sick of dating and dead-end relationships, but Rosebery is not about Pete. It's about our trip."

"Right, mates, Rosebery it is," Rhonda says as she bites into her meat-and-potato pie.

The following day we head south to Rosebery. We drive through the thick Tasmanian brush for hours, surprising wallabies on the road and catching glimpses of the rocky coast. There are no towns anywhere along the route. Then, suddenly at the bottom of a hill, numerous tin-roofed shacks appear lining both sides of the asphalt road. We're in Rosebery.

It's dusk when we arrive. Thirty cardboard-like buildings hang off the canyon walls. Gray with soot and skirted by slag piles, there's not a spot of color in the place. Mining towns are strictly utilitarian. Get the zinc out, make money, leave. There's nothing solid about this town, no street lights, no sidewalks, no brick houses. No curtains are in the windows, no flowers grow in the yards, and bare-bulb lighting adds to the bleak décor. This pre-fab town is just waiting for the zinc to play out so it can be folded up and moved to the next strike.

We inch the car through town navigating our way around burly men holding beers and yahooing at us. They're all unshaven and covered with dust. They range in size from small to huge, but all burst with muscles. At the only stoplight in town, a red-bearded man throws himself across our car hood and plants a drooly kiss on our windshield. Rhonda steps on the gas. He rolls off the hood, spilling beer from his schooner glass as he goes.

"We're going to spend the night here?" Haley yells. "Are you crazy?"

"Right, mate," Rhonda says, "We're the only bloody women in the place. Look at these wankers! Good job, Miss Rohver. What other travel suggestions do you have for the trip?"

Without answering, I push myself further into the back seat. We park the car at the, mine office, and go in.

"G'day, We're looking for Peter Fischer. He's working in the mine. Could you look him up for us?" I ask. Ian, the clerk at the desk, shuffles through a Rolodex and says,

"He's still in the mine on his shift but will be out in an hour. I'll flag him down on his way out. You can go to the Swan Pub. I'll tell him you're waiting there."

"Ta, you've been a big help." We find the bar, order some Cascade, Tasmania's favorite brew, and wait. Multiple drunken men approach us; Haley and Rhonda sit rigidly upright; casting dark looks my way and not saying a word.

Pete arrives, dirty-faced and carrying a lunch bucket. His smile gleams through the mud. He's more handsome than I remember.

"I'm so glad you decided to come! Three good-looking women asking for me at the mine office sure has upped my personal stock! You've made me the most popular bloke here!" he says wiping his hands on his jeans.

He stands in front of us and introduces himself. "Of course, I know you, Kris, I'm glad you decided to come. And you must be Haley with your black hair and Irish eyes and you must be Rhonda, the athlete." He says as he extends his hand. "I'm Pete. How was the trip? Did you see some amazing wildlife and scenery? I've only seen it from the bus window, so I imagine it must have been spectacular from the car," Pete continues excitedly exuding his endearingly boyish charm. I sigh with relief. He's here. I'm off the hook with my friends. Pete's even remembered our names.

Haley responds with her best smile, "Yeh, there was no one on the road, and the coast looked pretty rugged. Rhonda almost nailed a wallaby or two, but we made it,"

"I'm glad you didn't hit it, Rhonda, cause I haven't been here long enough to know how to make wallaby stew. I guess now that you've made me famous, I need to buy you another round." Pete heads to the bar bringing back four schooners of beer. "I promised Kris I'd find you a good place to sleep. I play chess with some university engineering interns who live in separate dorms near management up the hill away from the miner's quarters. They'd love to meet you. As you can see, the fairer sex is a rare species here. These Uni guys are real gentlemen. Unfortunately, there's only room for two with the Uni guys. It's the middle of the week, and everyone's here and working, so finding another room will be hard. Maybe Kris can stay in the miners' quarters? We each have our own small room and share a bathroom. It's small but OK. I'll give you my bed and sleep on the floor."

"That'll work for me," I answer. It's the 70s and "free" is a word we young, cheap travelers like. We regularly had young backpackers traveling through Australia sleeping on our apartment floor or sofa.

"Good on ya, mate," Rhonda says. She chugs her beer and looks at the exit. She's ready to leave.

Pete adds. "The mining blokes I work with in the mine aren't bad sorts. I've met a lot of farmers who are just trying to supplement their income to keep their farms. Others don't talk about their background, and I don't ask."

I glance around the bar and say, "I'm sure you're right." Although we've just recently called them wankers I happily agree.

Pete settles Rhonda and Haley in a room shared by two engineer interns. They willingly give up their beds and bunk in with friends in return for laughs and drinks with young women.

I end up staying in the mine's single men's quarters with Pete. Is it totally circumstance or planned? Maybe some of both. The thought of being alone with Pete overnight; interesting.

CHAPTER 3

Catch a Falling Star

"Follow me, I want to show you something," Pete says as he takes my hand. We go behind the shacks where the miners have put up Christmas lights surrounding their makeshift soccer field. It's three in the morning. He grips my hand tightly and then stops mid-field. "Look up." He grabs my other hand and pulling me around to face him. I look up. Millions of stars shine down from above. The gentle light reveals our faces. He leans down, kisses me, and wraps his arms around me, grounding me in a moment that feels infinite.

He slowly unwraps his arms, holds onto my hands, and we stand silently looking at each other. Silent night, holy night. All is calm. All is bright. Are we two strangers shaking off loneliness in a foreign land, far away from home or is this more?

A shooting star streaks across the sky. "Make a wish, Kris," Pete says.

"I wish, I wish this moment to last," I silently say to myself.

"What did you wish?" Pete asks.

"If I tell it won't come true," I say.

We follow the trail back to his room.

We sit on the edge of his bed holding hands. He kisses me softly and we lay back on the bed together. In the morning I awake still in his arms.

We meet up in the morning with Haley and Rhonda for breakfast. Pete offers us a mine tour, but Haley and Rhonda are eager to get to Hobart. I accept his tour offer and promise to meet them at the Hobart Hostel the next day.

To go down into the mine, you need a visitor pass and a guide. So we go to the main office, and Ian, a retired miner, agrees to take us down. There's a tin-roofed building next to the start of the narrow-gauge train going into the mine.

Peter goes inside his locker and grabs his coveralls. Then he picks up headlamps, hardhats, and battery packs for each of us. We climb into the open-sided electric train and head into the mountain.

Ian roars above the clatter of the tracks, "This here train takes us into the mine and out! It's our lifeline and our bread line 'cause it also takes the ore out. It's a pretty important piece of equipment,"

Halfway into the mine is a large circular elevator surrounded by chicken wire. It can hold thirty or more men at a time.

"Flip on your lantern, Missy. It gets darker than hell down here," Ian says. The elevator creaks downward. Every one hundred feet, tunnels go off in four directions, lit by strings of bulbs. Along the sides of the tunnels are channels for water and mud, with a train track running down the middle used to carry the ore out. At the fifteenth level, Ian drops off a pneumatic drill and we continue to go down to level twenty-four, near the bottom of the mine, where Pete usually works. The dampness and dark shake me. How far are we now away from air and light? It's farther than I want to be, but Pete and Ian banter lightheartedly.

"Yesterday, my mates and I were sitting in the stoop playing cards with our sweaty hardhats off, when a boulder crashed down in the middle of the table. You should have seen how quickly we all dove for our hardhats," Pete tells Ian.

"Yeah, those hardhats are a damn nuisance until you need them," Ian responds.

Pete looks at me and confesses, "I'm not really a true miner just an underground laborer, moving timbers and shoveling mud and ore. The real miners do the drilling and setting off charges."

We step into a hollowed-out stope (a stope is a room made by cutting out the ore.) It's cold, wet, and smells.

"This is where the miners figure out where to drill and plant the charges. Then they stuff the hole with a granular gelignite and explode the charge at the end of their shift. After the miners dig the ore out and as the level empties, they fill the space with a slurry of what's left after the ore is processed," Ian explains.

"Is it dangerous?" I ask.

"Well, Miss young sheila, in a word, yes. We're two thousand four hundred feet below our entrance tunnel; a lot of mountain on top of that. It's easy to have a collapse and for miners to be trapped." There's a strong coppery smell in the stone. "Sorry about the smell. Just last week a part of the ceiling fell, crushing a miner. It's from the dead man's blood. Not much ventilation here." Ian explains. I cover my nose and try not to hold onto the gory image in my head.

"In the side room off this tunnel, an old timer was working with gelignite and blew himself up. All we ever found of the dead miner was a boot." Ian laughs.

I secretly hope this is his last story. A sense of fear grips me. Is Pete safe? Will he be crushed or damaged by this dangerous job? Pete grabs my hand and squeezes it as my face tightens.

I'm glad when we get in the elevator and rise into daylight again. I can't believe Pete has described this as his best job.

Being so far down in the ground made me feel trapped and claustrophobic. I'm unsure I could spend eight hours underground without sun and fresh air. It scares me that he does. When I say this to him, he says, "The money's great."

"Thanks, Ian, for the tour. It was bonza! I love working here." Pete says as he shakes Ian's hand.

"You have fun, you two," Ian says, giving Pete a big wink.

Before I know it, three days have slipped by, and I'm still with Pete in the single men's quarters of the Rosebery mine. I never think about Haley and Rhonda worrying about me.

Pete works his shifts, and I read, sleep, and walk around the countryside. It's almost Christmas, so I decorate the room with handmade paper ornaments and sprigs of evergreens. We eat crackers, sardines, and whatever Pete can sneak out of the cafeteria. There's a shared bathroom, and I use the shared bathroom cautiously between shifts when nobody's around. I write poetry and put it up on the walls.

Raining all around me,

Moving me

Mooding me.

Now you all around me

Moving me

Mooding me.

I am without defenses to the weather

And your love.

We talk nonstop pouring out our twenty-three years of life to each other.

A Christmas package arrives from his mom, and we open it together. There's a Pendleton shirt, marzipan Christmas trees, some detective novels, and a long letter detailing all the family news. It's in chronological order starting with the oldest sibling, Pete, going down to Mary, the youngest of the seven children.

It's popular in the early 70s to be at war with your parents, but Pete can't stop telling me how much he loves his parents and how great his family is.

My parents are divorced. My older sisters are married with their own families. I'm sure they'll be no Christmas package for me in Sydney. Will anyone even miss me? Since my first year in college, I haven't returned to the home I grew up in, nor have I been with all my family during the holidays. My family is all around the country; Dallas, Milwaukee, Baltimore, and Alfred, New York. Pete's talk about his intact family makes me yearn for happier days with mine.

Pete takes the bus with me to Launceston and then to the airport. We haven't made any plans, but Pete says, "Maybe I'll quit this job and come to New Zealand to travel with you."

I smile and hope.

"Check poste restante (is a service operated by post offices by which letters and packages sent to you are kept at a particular post office until you collect them) in Auckland. I'll send you a letter if I'm coming," he adds.

As I'm walking to the plane, I look back. Pete is smiling and waving. He has on his sheepskin jacket, which is too short at the sleeves, his faded jeans, old hiking boots, and a wrinkled blue shirt that brings out the blue in his eyes. He looks beautiful to me. As we roll down the tarmac, I kiss my hand and spread it fanlike on the plane window. Will I ever see him again?

When I get to the train station in Melbourne, Haley and Rhonda are waiting. Rhonda cries out, "Crickey, mate! Where the 'ell 'ave ya been?"

"I know you're not going to believe this, but I think I've met the man I'm going to marry."

Meanwhile, Pete makes plans to leave the "best job of his life" and follow me to New Zealand. He gets an appointment with the mine manager to see if he can come back to his job. They say in the mining world that the mine manager doesn't report to God; God reports to the mine manager.

"I need to be gone a month or two. Is there any way you can hold my job open for me?" Pete asks.

"I only have one question for you, mate. Are you going to jail?" the manager asks.

"No, it's personal. I'm looking for a girl, but I don't even know her last name!"

"Good luck, young man. I hope ya find 'er."

Pete packs up his things and flies to New Zealand. He tramps around the North Island of New Zealand for two weeks without any luck finding me. He writes a letter home to let his family know where he's headed, why, and where to send his mail next.

From: Peter Fischer

Poste restante Christchurch, New Zealand

TO: Fischer Family

517 Ambridge Rd.

Calabasas, CA 91302

Hi everybody!

So much has been happening. I've gone to New Zealand to find a girl I met in Launceston. Her name is Kris, but I don't know her last name. I've searched for her for two weeks and am almost convinced she's not here. I have fallen completely head over heels in love. I hope I can find her.

Love,

Pete.

CHAPTER 4

Stars Align

Haley, Rhonda, and I return to Sydney to spend Christmas. Rhonda goes home to her mom in Mudgee. I cancel my date with Ian, the New South Wales Unie football player. I'm no longer interested.

Haley and I have a traditional English Christmas dinner at a friend's house. We sit packed around their dining room table with sweat dripping onto our plates of roasted turkey, parsnips, carrots, stuffing, gravy, bread sauce, and pigs in a blanket. A small fan in the corner gives us a bit of relief from the ninety-five-degree Australian summer. The Watsons serve us steaming bowls of Figgy pudding as we sit on the verandah in our shorts and tank tops. This rich brandy-infused Christmas cake is a delightful new taste for us Yanks. We thank the family for generously sharing their Christmas with us.

Other families can't replace your own, though. It's our first Christmas without family and we can't shake our sadness. It's so different here. In Sioux City, Iowa and Baltimore, Maryland it's cold and snowy. Everyone's Christmas differs, but we cherish our own traditions. Mostly we miss the warmth and love of being around people who've known us all your lives. At least we're together. We make three-minute calls to our families which costs us twenty Australian dollars per call. Haley barely has time to wish everyone in her large Irish Catholic family, "Merry Christmas." I call my mother in Baltimore and my father and stepmother in Belmont, New York.

"I'm still down. Maybe we should do something we wouldn't be able to do at home," I suggest.

"Let's go to the beach!" Haley says.

We head to the beach, where it's so hot the sand burns our feet and the shark flags are flying. No beach today.

"There's only one thing left to cure our homesickness, mate, drink," Haley says.

"Where? Won't every bar be closed on Christmas?" I reply.

"How about Sydney's International Airport?" Haley says.

"Good thinking, Haley. Let's go,"

The drinks are expensive, so we settle for Carleton beers. Lucky for us, though, we get a flat tire going out of the airport parking lot. The four parking attendants, who are about our age, bored and mad about working on Christmas, help us fix our tire. Then they invite us into the booth where somehow, they've gotten a hold of a large stash of liquor confiscated by customs. We wash away our homesickness with some strong, expensive grog.

Hungover, we fly to Auckland, New Zealand—the next day. The main hostel where we plan to stay is overbooked. Many young Australians are doing what we're doing—itchhiking around New Zealand for the Christmas/summer break. The hostel host sends us to a dodgy overflow hostel in a weird guy's house.

We decide to leave Auckland the next day rather than go back to that house. Before leaving I check the main post office for news from Pete. Unfortunately, there's no mail for me. Did he write to me or not? Is he coming or not? Has his letter not arrived yet? Maybe he doesn't feel as strongly about me as I do about him. Cloudy doubts swirl around in my mind.

Haley and I hitchhike around the North Island and end up in Wellington, a city at the Southern tip of New Zealand's North Island. It's New Year's Day and all the businesses are closed, so we sit in our hostel, sharing some biscuits and lemonade. We'll take the ferry connecting the North Island to the South Island of New Zealand tomorrow.

"What's going on, Rohver? You're no fun now! All these interesting guys from Canada and the US in our hostels, and you just read a book!" Haley says.

"I guess I'm not interested in dating anymore," I say.

"You're stuck on him. Aren't you?" Haley prods.

"No! I don't know if I'll ever see Pete again. Either he's still working in Rosebery, or he's continuing his world trip. I have no way to contact him, and he has no way to contact me. It wasn't just one of my flings. It'll take me a while to get over this guy. Maybe we'll meet again, but the odds are against it."

'You've always been good at loving and leaving them. This is not like you at all! So, get over it, Rohver!"

"I don't know, Haley. It's different this time. I never ran out of things to say to him. Even the silences were comfortable."

The next day we line up, tickets in hand, to board the ferry to the South Island. Unluckily, it's been raining for days. Overcast skies promise more rain and miserable hitchhiking.

Suddenly, I feel a tap on my shoulder. There behind me is Pete! After days of uncertainty, Pete's tap on my shoulder sends a rush of relief and joy through me. There he is, with his crooked smile and ill-fitting jacket, as if fate had hand-delivered him to me.

Hoisting his army pack higher up on his back, his two free arms wrap around me and crush me against his chest. "I thought I'd never find you." We embrace for what seems like forever until out of the corner of my eye, I see Haley, one hand on her hip, holding her cigarette glaring at us. Her voice jars us apart and we look in her direction.

"Hey, here's my ticket, Pete. I'm gonna hook up with a new friend we've met, Edna, and take the next ferry—two's company. Three's a crowd. I'm outta here. Have fun, you two." As Haley disappears into the crowd with her red kerchief fluttering, I realize my travels with her are ending, and a new chapter--with Pete--is just beginning.

Haley and I have been together ever since my junior year in college. This past year we've been each other's closest friend sharing our tears, happiness and secrets. We made a sacred vow to never put boyfriends ahead of our friendship, but now, here I am happily leaving her for Pete. There's resentment in her rigid posture and decision not to travel with us. A part of me wants to make Pete, Haley, and I, the three musketeers, but unconsciously, I know my life is shifting.

The ferry crosses the Cook Strait to Picton in three and a half hours. It's blustery and rough, so we sit inside during the trip. Most of the other backpackers head south to Christchurch once the ferry docks, but we decide to head west instead, taking the road less traveled.

"I hate to follow the crowd. I don't want to go where everyone else is going. I've heard the west coast of New Zealand is spectacular, so let's go that way," Pete says walking in the opposite direction from the other backpackers.

"OK, I'm game. Let's see where we end up," I say.

A farmer picks us up and leaves us fifteen minutes later in a small town with only a gas station and a grocery store. The hours slip by, and then the storekeeper and his wife cross to our side of the road and bring us a table, chairs, and afternoon tea.

"Welcome to New Zealand! We don't get too many travelers here. Where are you from, and where are you going?" the store owner asks.

"I'm from California, and Kris is from New York. We heard the Tasman Sea coastline is spectacular, so we thought we'd head in that direction, but there doesn't seem to be too many cars headed that way," Pete answers.

The storekeeper smiles and chuckles, "This road dead ends ten miles east of here. It's getting dark. I'm sure our neighbor, Trevor, won't mind you camping in his field."

So that's how our first day in New Zealand ends in a farmer's pasture.

Loud mooing outside our tent wakes us up before sunrise. Several cows are grazing nearby, so we pack up and return to the road. After stocking up on food at our favorite grocery store, we hitchhike back to Picton, buy a road map, and head down the "road most traveled" to Christchurch.

Since my summer vacation ends in three weeks, we race around the South Island, barely having time to take in the breathtaking fjords, mountains, bays, and beaches. First, we visit Christchurch, a small city with European architecture and an English feel. From there, we hitchhike to Dunedin, which precariously hangs off steep cliffs above the South Pacific.

The drivers who pick us up usually insist on us staying with them. A single mom offers us her guest room and feeds us dinner. When she has an errand to run, we babysit her two school-age children in a park by her house. As a teacher I love children, but I don't know how Pete feels about them.

"My name is Pete and this is Kris. Let me guess your names." Pete says as he points to the five-year-old boy. "You must be Mary!"

"No. That's a girl's name!" the little boy answers.

"Kathy? Abigail?" Pete guesses.

"No, silly I'm George," the boy says.

Pete then points at the eight-year-old girl. "Are you Frank? Harry? Ian?"

"No. I'm a girl. My name is Bethany," she says indignantly.

We hoist the kids on our backs for a chicken fight and end up in a pile together laughing. Pete is a natural with kids and knows how to have fun with them. I knew he had six younger brothers and sisters but seeing him play with these kids convinces me he'll be a good father.

Our next stop is Invercargill, where we end up riding with a priest who puts us up in the Catholic Boy's Boarding School where he works. The teenage boys flock around me asking me questions about the US and making fun of my American accent. It's rare for them to have a young woman in the house. New Zealand is a "small town country" where trust and friendliness are standard for the people. They willingly share their homes and lives with us. This is the beginning of the huge cast of characters we'll meet in our travels. Rarely, do the people of the world disappoint us.

From Invercargill we go South to Bluff, the nearby harbor. We take the ferry to Stewart Island and trek through a rainforest to camp on a beach. It's cold and rainy, but the black sand beach with granite rocks and crashing surf is worth the discomfort. We're barely able to sleep, but I don't complain. Pete's optimism and enthusiasm are contagious.

At the Franz Josef Fox Glacier, we have a dorm room to ourselves. In the morning we walk up to the first glacier we have seen to find its base littered with trash, dirt, and scree. It's a disappointment. The day is overcast and drizzly. As we look up at this wall of dirty snow, full of boulders and crevices, we have no desire to climb further and retreat to our two-bed "private" dorm room at the hostel.

We lie on bunk beds listening to "California Dreaming" and Don McLean's "American Pie" on Pete's small transistor radio. The songs remind us of places we know and American youth culture. We never run out of things to talk about and the lazy afternoon drifts on.

Pete explains why he never wrote to me. "I never learned your last name or Haley's" he confesses, "I searched all the hostel's log books for Kris and Haley, but both of you weren't there!"

"Well, that explains it! When you sign into a hostel, you use the name on your passport. My legal name is Elizabeth Kristina Rohver. Haley's real name is Norah Haley."

"Running into you in that WC queue was pure fate! Fate must be aligning our stars and lives!" Pete says.

As the lazy afternoon drifts into night, we realize how comfortable we are together. Our future travel plans start stretching out into the following year, the year after, and the year after that. Suddenly, we understand we're making plans for a lifetime, not just tomorrow's trip.

"Our plans are stretching pretty far in the future, Pete. Are we planning a lifetime together?" I ask.

As I say these words, I think. I'm not the girl who's dreamed about marriage and a big wedding. I've never said, "I want to get married". Even with Ron, where marriage seemed like the next logical step after two years of dating, I resisted getting engaged.

The emotion I'm feeling with Pete is different. There's a level of trust and comfort I've never experienced before with anyone. Although I've only known him for six weeks, I feel totally accepted by him. I'm not worried about what I say or do, he loves me for who I am and I love him as well.

"It looks that way." He leans over, kisses me, and ties a bit of yarn around my ring finger. I know the yarn on my finger will not convince my friends of our seriousness, but I wear it faithfully for months.

As I twirl the yarn on my finger, I think. What am I doing? Am I actually saying yes to marrying a guy with no job, no home, no car, and his immediate plans are more travel?

CHAPTER 5

Pete's Accident

It's January 14th. My new school year starts in two weeks so we hitchhike from the Franz Josef Glacier to Christchurch where I'll catch the plane back to Sydney. At the boarding gate, we say goodbye. "Here we are again. You're putting me on a plane, and parting! Are you sure you have my name this time? Don't forget me."

"No, Miss Rohver, I won't forget you. But, yes, Miss Rohver, I know your name and also have your phone number and address. You're not getting away from me again."

Pete grabs me tightly giving me a long, lingering kiss. "That'll have to last until I get off the cruise ship in February when it docks in Sydney. I think I'll return to Queenstown and hike the Routeburn Track before I leave.

"Bye, Pete. Be careful. I love you," I say as I head to the plane.

Once I'm back in Sydney, time speeds by, as getting ready for a new school year always does. There are new students to get to know, a new curriculum to learn, and a new classroom to be set up.

Haley, Rhonda, and I find a newly furnished apartment in Cronulla with two bedrooms, much cheaper than Haley's and my first apartment. We all need to start saving for our trip back to the US. Rhonda has decided to go with us.

After a few glasses of wine at night, Haley and Rhonda grill me about "your miner," as they call Pete.

"So, you're madly in love and engaged, but you have no man and no ring," Haley says as she looks at me skeptically.

"It's hard to explain, but I know he's the one, and I don't need a ring to prove it," I answer.

"One week with the bloke at the mine, you decide to marry him. But, really, Rohver, how naive are you?" Rhonda chimes in.

"I spent three weeks traveling around the South Island with him. Doesn't that count for something? Haven't you ever heard of 'love at first sight'?" I ask. They both laugh and roll their eyes.

"Hey, it's been two weeks with no letters or calls from the miner. Accept it, Rohver, It was a summer fling. We're your friends, and we don't want to see you get hurt," Haley says.

The next day an aerogram (thin, blue folded airmail stationary) arrives from Pete. I wave it in front of my friends triumphantly. "See, true love!"

Dear Miss Rohver,

After I left you, I decided to continue looking around the South Island. I got lucky and connected with a couple of Brits who were very pleasant. One of them had just come from New Guinea, where he had a job on a seismic crew out in the jungle. They wanted to hike the Routeburn, and with nothing better to do, I agreed to join them.

It's thirty-two kilometers long, beginning at the head of Wakapitu Lake and ending on the road from Te Anau to Milford Sound. We started in Queenstown, on the lake. We stayed one night in a huge hostel. We didn't want to pay the next night, so we slept on the beach but went into the hostel to make breakfast. We got busted by the hostel warden, who called us 'bludgers." It was a little embarrassing. Soon the boat will leave for the upper end of the lake, where we'll begin our trek.

I hope your school year is beginning well and you miss me as much as I miss you. I'd better pop this in the mailbox. The boat is boarding, but I'll write again soon. Try to keep Haley and Rhonda out of trouble.

Love, Pete

Shortly after the letter arrives, Pete calls. Calls are rare. Calling internationally in 1973 was prohibitively expensive. Three minutes cost you twenty dollars. When the phone rings, we are sitting around our dining room table doing lesson plans. I pick up the receiver.

The international operator asks, "Is this Kris Rohver? Will you accept charges from Pete Fischer?"

"Who? Pete Fischer?" I ask loudly enough for Haley and Rhonda to hear.

The operator repeats his name, "Pete Fischer."

There's a moment's pause, and then I laugh and say, "Yes, of course, I'll accept charges. Hi Pete!"

"Hi, Kris. It's great to hear your voice. I'm calling because I've been in a truck accident outside Te Anau."

"Oh, my God! Are you OK? A crash doesn't sound good."

"I'm pretty banged up with bruises on my back and shoulder, stitches in my head and neck, and a bad laceration on my left arm, but I'm alive."

"What happened?"

"It's a long story best told with a beer and a smoke. I'll tell you when I get back to Sydney. There will be no more trekking around New Zealand's Southern Alps for me. It'll only be a few more weeks before I'm docking in Sydney. See you soon. Love you. Bye."

"Bye, Pete! I love you too!" When I get off the phone, I break down in tears. "I can't believe I pretended I didn't know him when the operator asked if I'd accept the call!"

"Are you OK, mate?" Rhonda asks as she scans my face.

"No, Pete just called. He's badly hurt in a truck accident! He's in a small town in New Zealand. What kind of medical care can he get in a little town in the mountains? I'm worried what might happen!"

Rhonda rushes to me and hugs me. "No worries, mate. He'll be right. You'd be amazed at what they can do in those backcountry places. Sure, it's wop-wop, but they're amazingly self-reliant. He'll be good as new when you get him back."

As the date of Pete's arrival approaches, I wonder if he and I will feel the same way about each other. How will his accident have affected him? Will he get along with Haley and Rhonda? I'll be supporting him until his arm is rehabbed. How will that work? It'll be different than our vacation whirlwind, but I'm happy and excited about his coming to live with us.

When his cruise ship arrives in Sydney Harbor, I'm ready. I buy a nurse's cap at a uniform store, and through a friend who's a doctor, I get a blank prescription sheet on which I write in my worst scrawl. "Please release the patient, Peter Fischer, to the care of nursing sister, Kris Rohver."

Arriving at the dock, I wave my prescription in front of the ship's security and say, "I must immediately board the ship to take charge of my patient." I run up the gangplank and find Pete standing by the rail smiling as if to say, what are you doing, crazy woman? A large cast runs the length of his left arm, held by a sling. I take his backpack, and we march through immigration and customs ahead of all the other passengers. It's been over a month since we've last seen each other, but nothing has changed. We ignore everyone, but each other.

Later that night, Pete leans back on the sofa, takes a swig of beer, a drag off his cigarette, and begins his tale.

"After we got off the boat, we started the trek. There were huts along the way with bunk beds, outhouses, and kitchens. We'd wished then we'd left our tents and camping cookstove behind. This was luxury camping for all of us. It seemed strange, but we never saw another hiker the entire trip.

"What was the scenery like and how steep was the trail?" Rhonda wants to know.

It was typical South Island, lovely with soaring mountains, huge valleys, waterfalls, and jewel-like lakes. It wasn't too hard a hike, and we made good time."

"Did you see any wildlife?" Haley asks.

"We saw large gray parrots and flashes of red and yellow birds flying through the beech forest," Pete says. "We also saw lots of dead deer."

"What? Oh my gosh! Why? Where?" Haley queries.

"Deer are an introduced species so they have to cull the herds to preserve the native plants. It's not pretty to see. They shoot the deer from helicopters and then fly the carcasses out in big nets."

Pete continues, "When the two Brits and I finished the trail, we were on a gravel road connecting Te Anau and Milford Sound. My buddies decided to hitch to the Sound while I went the other way towards Te Anau to check into the youth hostel. They got picked up quickly, but I had a harder time. There was almost no traffic. Then, along came an old pickup truck with a flatbed trailer on the back. Live lobsters in gunny sacks were piled up in the back."

One minute I'm joking and laughing with the fishermen and then the truck skids on the gravel road, tipping over as by body slams into the cab's frame. The world went dark. When I came to, I was on my hands and knees, blood pooling around me.

I make out three men standing over me. "Don't stand up!" one says and then they carry me to the bus and lie me on the bus's backseat. I feel the bus turn, and accelerate then stop abruptly. I hear the bus driver yell, "Does anyone know first aid?" There's no answer so he speeds out of the store's parking lot. I see blood dripping onto the bus floor. Is this my blood, I wonder? From there it's a bit foggy. When I wake up there's a man and woman bending over me. "What happened? Where am I?

"You're in Te Anua. You've been in a truck accident. My wife and I have been working on you quite a while. Your forearm was cut, severing several tendons to two of your fingers. I reconnected them which was a bit tricky and stitched up deep gashes in your neck and head."

"The driver of a tourist bus going the opposite way saw the rolled over truck, the lobsters strewn on the road, you bleeding, and the two young New Zealanders waving frantically at him. He stopped the bus, and brought you here," the doctor explains and asks, "By the way where are you staying?"

"The youth hostel here in town. I'm sure I'll be fine for a couple of days."

"Oh no, son, that isn't happening. Let me call our friend."

When the friend arrives, I turn to the doctor and his wife and say, "I can't thank you enough. How much do I owe you?"

The doctor answers, "Based on the look of you, you don't have too much."

To which I replied, "No, but whatever I have is yours if you want it."

He laughs and said, "I'll be happy if you send me a bottle of whiskey when you get to Sydney and let me know how your arm is doing."

"Their friend sets up her spare bedroom for me and asks if there's anything else she can do. I tell her about my two Brit friends coming to the youth hostel later that night and whether she would mind calling them to let them know I'm OK.

"She drives to the youth hostel, picks up my friends, and puts us all up for a few days. She's a typical wonderful Kiwi! Amazing people!"

I was in a cast on my arm, so I hung around until the ship left from Christchurch to Sydney with no other mishaps."

Rhonda exclaims, "I told you, Kris, he'd be more than right in the backcountry. Those Kiwis are fair dinkum!"

"I'm glad I never knew how badly you were hurt when you called," I say as I lightly touch Pete's shoulder.

The next day Pete goes to the Sydney orthopedic surgeon whom the doctor in New Zealand had recommended to rehab his arm.

"Pete, you said you had this done in Te Anua, New Zealand, right? Isn't that a pretty small place?" the doctor asks.

"It's got about two-thousand people; why?"

"Most doctors wouldn't have the skill to connect these tendons in the arm to your hand well enough for you to regain full use of your hand again, but this doctor did. You are one lucky guy." the surgeon says.

"I guess I'd better send him that whiskey I promised him."

"You'd better!"

Pete goes through several weeks of painful physical therapy but eventually regains full use of his left hand and arm. Today all that remains of the accident is a long scar across his left forearm. Pete was patched up by a fly-fishing Canadian doctor in a remote clinic—a lifeline in the middle of nowhere.

CHAPTER 6

Wedding Down Under
Dec. 15, 1973

On returning to Sydney, Pete moves in with Haley, Rhonda, and me in a small apartment in Cronulla. This tourist-suburban city is at the end of the local train line south of Sydney. It boasts a mile long beach that curves along the west side of town. Haley and I had lived there last year in a swanky apartment overlooking the ocean, but we've downgraded to a smaller, cheaper, apartment not on the beach.

A few weeks later, when we're in bed, Pete puts his arm around me and pulls me to his side. "There's something we need to figure out. Do we want to get married here in Australia before our trip or when we get home?"

In New Zealand, Pete wrapped yarn around my ring finger, but he never mentioned marriage. Although we'd talked about future trips and traveling, I wanted a real proposal, romantically, thought through. I didn't get it. I didn't know then, that Pete, through all our years of marriage, would never be romantic.

"This is so practical and unromantic. I don't know what to say! Are you asking me to marry you or making a 'to-do' list?"

"Well, are we getting married here or there?"

Tears stream down my face as I turn my back to him and leave the question unanswered.

Urged on by Haley and Rhonda who are always up for a party, Pete agrees to have an engagement breakfast party with pancakes and beer to compensate for his less-than-well-thought-out marriage proposal. We set the wedding date for the day after school lets out, December 15th.

For the engagement party, everyone invites their friends, and the party spills out into the hall. The festivities go late into the night with several beer and food runs. It's so out of control we get evicted from the apartment the next day.

We find a better, bigger apartment and ask Kathy, a teaching friend, to join us. Now Pete is living in a one-bathroom apartment with four women. Unfortunately, what he always thought would be a dream come true isn't. Navigating hanging pantyhose in the bathroom, having the phone in continual use, and helping hungover girls recover after partying makes him decide to look for a new place for the two of us.

In Bundeena is a cottage. It's a fifteen-minute ferry ride to Cronulla but an hour's drive through the Royal National Park south of Sydney. Lots of artists live in Bundeena because it's so cheap. The town came into existence during the Great Depression and still squats illegally at the tip of the Royal National Park.

The house hangs off a cliff overlooking a river estuary going into the Pacific. Windows line the entire front of the house; it's lovely but cold and hard to get to from my school. To get to my job at the intermediate school in Gymea, I have to walk into Bundeena, take the ferry, then take the train to Gymea and walk to my school, an hour and a half trip each way.

Wedding Down Under Dec. 15, 1973

Kevin Whitley and Robert Thomas share the cottage with us. They're two Brits with different accents. Kevin is a car mechanic and x-motorcycle sidecar racer from Birmingham, and Robert is a farmer, now world traveler, from Romsley, Halesowen in Worcestershire.

During the months we live with them, they barely speak to each other and only seem to share their love of the Goons Show on the radio. They sit howling while Pete and I listen, wondering what's so funny.

The bachelors have definite ideas about weddings. Kevin is a welder and is excitedly planning the bachelor party, where he threatens to attach a ball and chain to Pete's ankle. I have visions of Pete clanking down the aisle at our wedding. I put an instant stop to Kevin's plan.

Traveler Robert is trying to convince us to wait until we get to Bali to get married. "Imagine the wedding party decked in bright silk and fragrant flowers, riding to the Buddhist temple on elephants with silver and gold headpieces! Lovely Balinese women in bright sarongs tossing petals in your path. Now that would be a wedding!"

"But Robert, we're not Buddhists, and would a marriage in Bali be recognized in the US?" I question.

"No, it's probably not legal, but what a spectacle!"

"Sorry, Robert, that's a big, fat N-O from this bride-to-be. Any more ideas, gentlemen?"

Because our marriage will be halfway around the world, we decide to make our wedding announcement memorable. Our US friends and family won't be able to come, but we want them to at least see us and our new other half.

We rent 19th-century costumes and pose around Sydney. For our wedding announcement, we pick a picture of us in front of the old sandstone pub/hotel behind our cottage. Our Bundeena landlords have generously loaned us their unrestored pub for our wedding reception and it turns out to be the best picture of both of us.

Dressed in our fancy costumes we smile out from the photo to our friends back in the US giving them our good news and asking for their well wishes for our December 15th wedding. We take the pictures two months before the wedding because we can't afford airmail and send them by sea mail to get to the US on time.

We plan to get married in the local Methodist Church in Bundeena. Pete is Lutheran, and I'm Catholic, so we decide to go somewhere in between with a Methodist ceremony. The church is a small wood-framed building with a famous stained-glass window done by a Sydney artist, featured in an article in the Sydney Morning Herald. It's an unpretentious building but has rustic charm.

We are required to take a pre-nuptial class before the wedding. I'm sure the minister isn't giving this marriage much of a chance. There's nothing ordinary about it; no engagement ring, no family involved, and no possessions.

We remind our Australian guests not to bring presents. They'll be no room in our packs for them as we trek homeward overland following the "Hippie Trail". We end up receiving telegrams, money, and well wishes on our wedding day.

Only Peter's parents, Steen and Mary Anne, travel to the wedding. They arrive quite unexpectedly on December 11th, a day after we'd been planning for them to arrive. When you cross the International Dateline, you lose a day and go forward a day. They had given us the wrong day for their arrival.

Wedding Down Under Dec. 15, 1973

For the past three months we'd decided not to let them know we're living together. I am a good Catholic girl and Steen made it clear to Pete early in his college career there would be no living with any girlfriends. When he did live with his girlfriend, Felicia, he banned Pete from the family.

At six AM there's a knock at our door in Bundeena and there standing on our backsteps are Steen and Mary Anne, Pete's parents.

They told us they'd be arriving on December 10th. They'd forgotten about crossing the International Dateline so they arrive December 11th a day later than expected. Out of respect for their feelings I had planned to leave Bundeena and stay with my friends in Cronulla, As I answer the door in my pajamas, they realize Pete and I have been living together. Our secret is out. I stand there silently. My face turns red and a crooked smile spreads across my face.

"Pete, your parents are here. Sorry, we were expecting you yesterday," I say as I usher them into our living room.

"Yeh. We forgot about the International Dateline and gave you the wrong day, so here we are," Steen exclaims! Pete comes out of the bedroom, rubbing his eyes and nervously looking at his dad. They are happy to see us. Steen never mentions our living arrangements.

(How quickly mores and attitudes change! On returning home a year later, Pete discovers three of his younger brothers and sisters living with their boyfriends and girlfriends without a word from their dad.).

Before I leave for work, Mary Anne gives me a cross Steen, Pete's Dad, had given to her when he proposed to her twenty-five years ago. As she hands me the gold Dansk cross, her eyes brim with tears.

It was a piece of her history and now it will become part of mine. I finger the dent in the cross, place it around my neck and tell her I will wear it at my wedding and cherish it forever.

Gymea Primary School's end-of-the-year musical is Thursday night December 12th. My third graders are singing aboriginal songs. Before the program starts, I hand out spears to my third-grade boys. They declare war on each other, and I declare war on them, grabbing spears and screaming at them. From the corner of my eye, I see Pete and his dad watching us through the classroom window. I imagine Steen wondering if his soon-to-be daughter-in-law will be a good mother for his future grandchildren. I cringe knowing I've been caught in a wild banshee moment and hope I'll have a chance to show my kinder, gentler self.

On the last day of school, I clean up my classroom. I have lost and found my purse at least three times as I go through my long wedding to-do list.

I head home in roommate Kevin's borrowed Mini Moke. On my way home, I hit a wallaby—a ridiculous heart-wrenching accident that somehow fits the madness of planning a wedding, ending a school year, and starting a worldwide trip.

Saturday morning, Haley and I plan to drive to the Sydney flower market and buy flowers for the wedding, but I forget our planned trip. With no transportation, Haley steals flowers from people's gardens to decorate the trellis at the entrance to the reception and our bouquets. Haley doesn't get mad at me, because she's a good friend and knows how overwhelmed I am. Ever since we moved out of the apartment, I've barely seen her. I no longer hit the pub on Friday nights and only occasionally stop by the apartment on my way home from school. Pete has taken the larger chunk of my time.

Saturday night is the wedding dress rehearsal at the church. We have written our own vows omitting the word obey. It's a control word we don't like.

All goes well, except Pete hasn't memorized his vows and stumbles through his part. He's waited until the night before the wedding to write and learn his vows. Pete is a "last-minute-guy" who hates to plan and loves events to evolve spontaneously. I'm just the opposite. I throw him my annoyed teacher look, but it's too happy a night to spoil.

Friends and family head to the local RSL (Return Service League) in Bundeena. These are veterans' clubs where the food is cheap, the music is lively, and the gambling is hot and heavy.

My future mother-in-law, Mary Anne, tries her luck at the slot machines for the first time in her life. Within a minute, she's hit the jackpot! Coins pour out of the machine.

Pete runs to his mom and lifts her up excitedly. "Mom, you won the jackpot! Beginners' luck!"

"Wow! Mary Anne, you won!" I scream.

The regulars eye her suspiciously. How did this rookie Yank, who doesn't even know how to pull down the arm, win? I'm afraid of what these sweet old ladies and men will do. Gambling on the one-armed bandits, turns them into feral predators who viciously attack anyone who messes with their machines. I'm relieved when they simply smile and congratulate her on her luck.

Their congratulations turn to disbelief as Mary Anne pulls the arm down again. By pulling the arm again she's lost half the jackpot, The casino workers now can't give her the rest of her winnings. Her naivete is charming to me.

Since the wedding and reception are the day after the last day of school, December 15th, the kind teachers at my school, Gymea, and Haley's school, Gymea North, put on our reception. It's both an end of year party for them and our reception. They bring all the food, set up the barbecue, and tap a keg of beer. The librarian from my school makes a wedding cake which replicates my wedding hat complete with a chocolate sash and sugary Australian flowers decorating its brim. These Aussies have gone out of their way to make sure we get a memorable wedding in our adopted country.

Our poet friend, Phil Roberts, reads an original marriage poem for us, our pianist friend plays music from Bernstein's "Mass," and another friend sings Bernstein's "Sing God a Simple Song." Then, finally, we walk out of the church to the Sandpiper's tune, "Come Saturday Morning, I'm going away with my friend."

My wedding reception is a blur. I have a wide-brimmed hat with Australian flowers and a silky, brown-colored ribbon. It costs more than the beige boat-necked blouse and long skirt I wear as my wedding dress. The new gold cross sparkles. Everyone's having a good time. I feel pretty, happy, and loved.

Pete buys a gray suit with bell-bottom trousers and a white shirt. He also wears a tie but instantly takes it off once the party starts. It's a good party, and the day is warm and sunny.

Midway through the party, our guests climb down the hill and swim. It's the official beginning of summer in Australia. Extra beer runs to the RSL Club keep the party going well into the night. I am now Mrs. Fischer, going away with my friend.

Wedding Down Under Dec. 15, 1973

Pete's parents, thoughtfully, book nearby motel rooms for all of us. Mary Anne gives me a sexy negligee for my wedding night. She's thought of everything to welcome me warmly into the family.

The next morning, we say our good byes. Pete has been gone for two years. One week isn't long enough to catch up. Many things are left unsaid. Now as a mother myself, I feel Mary Anne's sad-joy watching her child get married and start his own life separate from them. She wipes tears from her face and hugs Pete for five long minutes. "Take care of each other," she says and embraces us both. They will only hear from us through letters for the next twelve months.

"Take care of yourself, son. Remember, if it ever gets too tough or dangerous, fly home," Steen advises as he shakes Pete's hand.

"Pop, I have $500 in a bank account Mom set up for me for just such an emergency. Besides, I know some karate. Nothing is going to happen to us." Steen turns away too abruptly motioning Mary Anne to get in the car and they drive away. We're on our own again. The comforting blanket of caring family is gone.

CHAPTER 7

Crocodiles and Floods

It's been a week since our marriage in Bundeena and we're on the road headed to Townsend, Queensland. Our overland trip has begun. We plan to travel through Queensland and then to Darwin in Kevin's tubby, Mini Moke. The word moke is an archaic term for mule, and our "mule" is doing the best it can with all four of us plus our gear. Kevin, Pete's best man, is driving. Pete sits in the passenger seat and Haley, my maid of honor, and I are squeezed in the back. Behind us we're pulling a small trailer that Kevin welded together to carry all our camping gear and backpacks.

As the car sputters to a stop, Haley groans, "Another breakdown? At this rate, we'll never see Darwin before Christmas!" Kevin mutters something about miracles and mechanics, but all I can think about is when it will stop raining.

Our wedding party is hating each other more with each passing day. Haley is a buxom, outspoken, feminist from Iowa. Kevin is a short ex-motorcycle side-car racer and a confirmed, non-nonsense, bachelor from Birmingham.

For the past few days, we've been unable to camp and have holed up in small trailers in caravan parks. The close quarters are not helping the mood. To add to our discomfort the Moke's canvas sides can't keep the drenching rain out.

As we pass through Guthalungra, the traffic slows until at Inkerman, it stops altogether. "What the bloody hell is this?" Kevin spits out.

"Let me go and see." Pete volunteers. He walks down the long line of vehicles and returns fifteen minutes later. "The bridge at the Murrumbidgee River has been washed out. We'll just have to wait. The good news is that I saw our friends, Deide and Will, further up the queue." Slowly the cars between us and Will's car turn around and return to Inkerman.

Will pulls out a gallon of wine to share. In the fading daylight we see cars being towed across the river by an enterprising farmer in his tractor for twenty dollars each. The wine improves our moods and gives us false courage.

"I'm not spending that much money for a three-minute tow!" Pete says.

Kevin replies, "Look, this saltwater creek doesn't look too deep. Maybe we don't need to be towed. The moke is four-wheel drive. I bet we can make it across without the tractor!"

"Do you think there are crocs in the creek?" I ask.

"Very likely," Kevin replies, "but vehicle height is a bigger problem."

"You don't have much clearance in the moke, but our Hillman does," Deide observes. "We can go first and test the depth of the water."

"If I spend one more night on a pub floor or in a small trailer in a caravan park, I'll swim across this creek," Haley says.

"Think you can outswim a saltwater croc?" I ask.

"No, I'll just push this piece of tin in its maw before it can get me." Haley says.

Kevin groans and grits his teeth.

"I guess that's a yes vote from you, Haley," I check.

"You bet it is!" Haley answers.

"I've heard these saltwater creeks in Queensland are full of crocs," Pete warns.

"You really didn't say that, Pete?" Haley and I say in unison.

"The flood water wasn't enough? So now we can also worry about razor-sharp crocodile teeth!" Haley says glaring at Pete.

"Are we doing this?" Will drunkenly yells out. He's answered with five yeses.

We watch the tractor pull a few more cars up the opposite bank trying to gauge how fast the water's moving and how deep it might be. As night falls headlights gleam on the muddy, fast-moving water. I look for croc's eyes reflecting the lights and don't see any.

The Hillman goes first, with five of us pushing from behind and Kevin driving. As we enter the water, I realize how fast it's moving. The muddy bottom squishes around my sneakers as I push against the car's back bumper. There's no stopping until we're up on the opposite bank. We cross back across the creek. I again scan for hungry croc eyes.

The muddy water continues to churn, hiding who-knows-what below; crocodiles, dead bodies, eels? Kevin guns the motor, puts it in first gear, and we grip anywhere we can. We hesitate until Kevin yells. "Full speed ahead," and the moke plunges into the torrent. There's an anxious moment when the engine sounds like it'll stall, and water streams into the cab.

"Put your backs into it, mates!" Kevin calls out. We give it an adrenaline push, and the moke climbs up the opposite bank with a jerk and a cough. Muddy water streams off us and the jeep. On both sides of the bank, the crowd cheers! We're the first ones across without paying the farmer, and everyone likes that.

CHAPTER 8

How Deep is Your Love?

We had conquered the rumored crocodile-infested waters of the river, been shouted free schooners of beer for our bravery, and now we are on our way to Townsend and the Great Barrier Reef.

"Crickey! Have you seen the paper? All this rain has flooded the Northern Territory. We'll never get to Darwin overland." Haley loudly says from the back seat of the moke. She thrusts a picture of an outback Australian cowboy, standing at a pub in Darwin, knee-deep in muddy water, drinking a beer with the caption. "It'll be right, mate!"

"Gotta love that Aussie spirit. We still have two weeks before we head west across the outback. That dry land will suck up all this water in no time, don't worry, Haley." I tell her, "We'll make it. Don't you want to see Townsend and Cairns? The sun is out! Be happy!"

"I hope you're right. In March I'm meeting Rhonda, Helen, Kathy, and Carmen in Darwin for the flight to Portuguese Timor to begin our trip. I can't miss them," Haley says but then laughs loudly and says, "You never know what's going to happen when you're traveling, do you?"

Once we settle into the Townsend hostel, we sign up immediately for a snorkeling trip to Green Island in the middle of the Barrier

Reef. We'll get a ship ride out to the island and snorkeling equipment for the day. We pack a lunch and board a large ferry with sixty other people headed toward the reef. From the bow, we watch sea snakes and porpoises streaming past the bow. The deep blue Pacific slowly changes to a translucent aquamarine, and we see an island of white sand beaches and palm trees ahead. We're out of Sydney's dry, Mediterranean climate and in the tropics now.

Several times during the trip, the PA system blasts a warning, "The ship leaves promptly at three back to Cairns! There will be three warning blasts on the ship's horn, and then we leave for the mainland! If you're not on board, you will be stranded on the island, so please be prompt in getting back to the ship. Did they need to keep repeating this? Who'd be foolish enough to miss the boat?

On docking, we run for the beach and plunge in the water. Beneath us is a world we've never imagined. The creatures we see are bright pinks, oranges, and purples. Giant clams with large purple lips open wide when we prod them with a stick. Convict-like clown fish flit around us in their white-and-orange-striped jumpsuits. Large forests of salmon-colored fan coral spread across the ocean floor.

While snorkeling we stay near each other. When Pete stands up waist-deep in water, I follow his lead. "This place is amazing! It's a wonder of the world. I haven't ever seen anything like it. So many fish and so many brilliant colors. This trip to Green Island is the best thirty dollars I've ever spent."

"I agree. Amazing!" I say as we dip back down under the waves. Fingers of purple and yellow anemones sway with the waves, hoping to catch tiny fish and jellyfish. Sea stars grip the rocks. Bi-lipped parrotfish curiously approach our snorkel masks and then make a quick retreat. Seahorses dance, bobbing up and down in the seagrass and lettuce. Sponges spiral upward toward the light. Blacktip reef sharks dart in and out between the brain coral. Crabs scuttle along the sandy bottom, moving sidewise with their stalked eyes protruding from their shells like mini-subs. In the crevices, dark-colored eels hide, waiting to snap up a fish with their razor-sharp teeth. Giant-sized groupers, snapper, and parrotfish slide past, shimmering in the light. It's a colorful dream world with muffled sounds and muted light.

We head in different directions and almost immediately, I lose sight of Pete. He's entranced and mesmerized by the wonders of the Great Barrier Reef. For a moment I see him come up near the island's south end, and then disappear again into the reef lagoon.

Haley and I decide to explore the cool, shady rainforest ringing the beach edge. It's cool, moist, and green, a welcome respite from the intense midday sun. When we return to the beach, people congregate at the end of the pier near the pilings, where several palm trees provide shade. The sun has moved from directly overhead to mid-horizon. It must be almost three, but we have no cell phones or watches to confirm this. Once on the ship, I scan the crowd for Pete.

When I see Kevin, I ask, "Have you seen Pete?"

"No, I'm as white as an English mechanic, so I got off the beach after an hour. I left him poking around in the water and found a resort on the other side of the island with a pub. Nothing like a schooner of beer and lots of sheilas in bikinis to watch!"

"So, you decided, bikinis over one of the Seven Wonders of the World, the Great Barrier Reef. Kevin?"

He nods yes, boozily. Just then, we hear three loud blasts from the ship's wheelhouse.

"We'll be leaving in five minutes. Make sure you have all your belongings," the ship's loudspeaker booms.

I scan the beach, water, and ship's decks for Pete. I search the ship's decks again. He's not there. I race up to the bridge and pound on the cabin door. They open it, and I rush in wild-eyed and scream,

"Don't leave! My husband's missing!"

"He's aw right, Missy," the man holding the steering wheel says as he winks at me. "I don't see any dead blokes floating in the ocean."

"I'm not a missy. I'm a Mrs," I say, holding up my left hand with its new gold wedding ring.

"I'd like to help you, Mrs., but I gotta get this here boat through these reefs and back to Townsend before dark. I'll give her three more blasts, and then we leave."

I scan the beach for a silhouette, but the horizon is empty. The horn blasts again. I race down to the gangplank and grab Haley. "What should I do? Should I stay or go?"

"Listen, friend. This isn't my call. Will you stand by your man or be sensible and come back to Townsend with us?" I see the deckhands untying the bow and stern lines. "Better decide quickly, Kris. This ship is leaving."

"Wait, I'm getting off!" I shout to the deckhands. I bound for the gangplank and hop onto the pier just as the boat pulls away. I walk slowly down to the pilings at the end of the dock. I'm not sure whether I should be worried or angry. If he's had an accident or hurt himself, I should be there for him, but if he lost track of time, maybe I should have left him. Although there was no 'obey' in our wedding vows, we had promised to support and care for each other forever. Staying here with him; my new married status.

Fifteen minutes later, a grinning Pete is standing in front of me. His grin is infectious, even as I fume about being stranded. "Relax," he says, "We'll sleep under the stars—it's an adventure." He gives me a salty hug, and I will myself not to cry.

Just then three Aussies in khaki shorts, black tank tops, wide-brimmed outback hats, and flip-flops walk by. "G'days" ring out between us as they head toward a motorboat tied up along the pier.

"Bleu, give me the keys to the boat," the tallest man yells.

"Here you go, mate. Catch 'em!" His red-haired friend throws the keys with one hand while taking another swig from his Swan beer. The keys arch high in the air, sparkling for a moment above the water, then disappear underneath the dock.

"Waco Bleu! Bloody hell, mate! How are we gonna get back, now?" the third man chimes in and leans drunkenly on the pier railing. Pete runs toward them, waving his snorkel mask.

"Hey, guys. If I dive in and find the keys, could you give my wife and me a ride back to Townsend?"

"Sure thing, Yank!" the blond answers.

Pete dives in among the pilings several times but can't find the keys. The men, Pete now included, decide to hot-wire the boat. After a few tries, it roars to life, and we end up back in Townsend before Haley and Kevin get there.

While on the docks and after a few congratulatory beers with the motorboat men, we learn the blond man owns a Chrysler dealership in Cairns and knows most of the fishermen in town. We also learn that there's regular boat traffic between Cairns and the Northernmost point of Australia, Thursday Island. They mention a coastal trader called the Safari owned by Captain Webb, who sometimes takes passengers with him on his trips through the reef. Since flooding in the Northern Territory has blocked the overland route to Darwin, we've been worried about how we'll cheaply get to Bali. Maybe we've found a way.

We leave for Cairns the next morning having experienced the Great Barrier Reef at its best. The colors, abundant sea life, and variety of fish that we see on the reef in 1974 will never be duplicated again. The evil Crown-of-Thorns Starfish takes over the reef ten years later destroying much of the coral. Rising sea temperatures also hurt the eco-system. The Great Barrier Reef is memorably spectacular, no other coral reef we see in the future will surpass it.

CHAPTER 9

Thursday Island University: Beer, Women, Shrimp

It's two in the morning, and I'm sitting in the wheelhouse of the Safari. Captain Webb recites Henry Lawson's poems and nurses his third cup of coffee. The instrument panel glows in soft purples and yellows. The bow throws up a fluorescent froth. The moonless night wraps us in a tropical blanket.

As Les drones on, I think back to our last days together with friends in Cairns. We go on a reef fishing trip, and using only lines, catch giant fish which we fry—feeding the entire hostel. It's our last night. We head our separate ways. Haley flies to Darwin, Kevin goes back to Sydney, Deide and Will fly to New Guinea, and we go to Thursday Island. Mother Nature has upended our plans. The trip is taking us whether we like it or not.

Pete is sleeping in one of the bunks below deck, Tomorrow he'll help Captain Webb (Les) unload his cargo. Les's slightly slurry Queensland brogue almost puts me to sleep when the radio crackles.

"What the hell is this? It's two in the morning! Who could be calling me now?" He pushes the talk button saying, "Copy. This is Safari."

Static breaks up the message. There's a pause then the operator on the other end says,

"Brother dead."

Pause.

"What? Who?"

Pause.

"Dead, your brother."

There's more crackling, and Captain Webb cradles the receiver anxiously in his hand.

"Skydiving accident!"

Pause.

"How?"

Pause

"Chute didn't open."

The radio sputters, hisses, and then goes silent. Captain Webb pushes down hard on the transmission button, but the connection is lost. He stares unbelievingly at the receiver.

He drops his head, holds the receiver to his chest, and closes his eyes. "I can't believe it. My brother's dead? How could that happen? How could that happen? He's made over 200 successful jumps."

I softly touch Les's shoulder. "That's tough. How old is he?"

"He just turned forty-five." We sit in silence for several minutes. "Sorry, time to call it a night. I'll use George's phone when I drop off his supplies tomorrow. George has a radio tower so the transmission will be better."

We find out that George is an Australian WWII vet suffering from PTSD who lives as a hermit on this coral island which was once a military installation. The bunkers and radio tower still line the atolls shoreline. When the Safari comes into view, George is out in front of his thatched hut waving his arms in welcome. He's a skinny, tanned, elder man in khaki shorts, a bush hat, and no shirt. His hair is a bleached tuft that curls around his ears. We jump off onto the rickety dock. Captain Webb introduces Pete, and me to George and then, with a catch in his throat, asks to use George's radio.

"I got a message last night that my younger brother died, and I need to contact my family. The radio transmission broke up last night, so I'm anxious to learn the details."

Captain Webb strides off to George's hut, and George takes us to a table and chairs set up on the beach with biscuits and fruit drinks.

"Where are you two from? You have Yankee accents," George says.

"I'm from New York, and Pete's from California, but we've been living in Sydney for the past two years," I answer.

"You're a long way from home, youngsters!" George comments.

"So are you. How do you like living out here? It must get lonely," I say.

"It does. It does, but I have my cockatiel, Ting, and my cat, Goolong. Say, do you have any paperback books you might want to exchange?" We trade books and talk some more.

"How did you come to live out here alone?" Pete asks.

"Got tired of all the wankers and wackiness of the city. It's quiet and calm out here. Les, the Captain, brings me supplies from the mainland. I fish, grow a little garden, and read."

Les comes back to join us on the beach. His shoulders are stooped, and he looks tired. Then, wiping his hand over his face, he smiles shakily and says, "OK, kids, we need to get going. Tide never waits for any man. Bye, George, thanks for the use of your phone."

As we climb back onto the Safari, George says, "Sorry for your loss, Les." We wave goodbye to George. Captain Webb stands silently at the wheel, starts the engine and heads north zigzagging through the reef. At dusk he anchors the boat in a small bay. "I'm beat. We'll sleep here tonight. There's no point in getting to Thursday Island tonight; the customs guys only work in the morning." I give Les a long hug before Pete and I go down to the bunks.

Early the following day, we arrive at the Thursday Island dock. It's a paradise with palm trees, aqua-colored waters, and Torres Island Strait natives. Two of Her Majesty's crisply, uniformed customs officers lead us into the colonial-looking customs house, the only non-tin shed in eyeshot. Pete unloads the ship with Les; we say goodbye, and head out to explore the small island.

As we walk through town, the Torres Island Strait females eye Pete hungrily. An overweight woman in a loose-fitting, red-flowered housedress gives him the infamous "Thursday Island handshake," grabbing his penis and squeezing it. Unfortunately, the Torres Island Strait natives are a terrible mix of Aboriginal, Polynesian, and New Guinea natives. The combination is unappealing, but their smiles and playful nature are infectious. Pete's face registers shock, but he smiles back at the woman.

In town, we meet Ken Daniels, a South African traveler who has just taken the overland route from Amsterdam through Asia to Australia. The same route we'll take, but from the opposite direction.

His tales of adventure and danger in the east excite and scare us. He helps us find a place to stay on the island. It's a tin shack with a sink and an outhouse. At least it's cheap.

He also warns us, too late, about the aggressive Torres Island women. He tells us that he went against his apartheid upbringing and slept with a native woman, which caused him to break out in a rash all over his body. He went to the local hospital, fearing the worse, but discovered it was just an allergic reaction, not a sexually-transmitted disease. My old boyfriend, Ron, was black. Ken's racism bothers me, but I don't say anything.

Walking around, we meet another Ken. Ken Johnson is a young Australian artist from Surrey Hills. He's here doing a series of sketches of the tropics.

After getting a place to stay, we head back to the docks to find a boat to take us to Port Moresby, but no one seems to be going any time soon. We are stuck, for now, on Thursday Island.

That night we go to the open-air theater with eight rows of wooden benches in front of a pull-up slideshow screen. "Charlie" and another old jungle movie with crocodiles are playing. The setting is heavenly. The stars are the theater's ceiling. Silhouetted palms surround us, and a soft breeze smelling of honeysuckle fills the air. Flying foxes swoop through the night, appearing briefly onscreen as dark-winged avengers; they're a unique addition to movies on Thursday Island.

In our search for a way to get to New Guinea, we meet Kazuie Matsumoto. Kazuie, the pearl farmer is a man of contrasts—flamenco guitarist by night, karate master by day. His stories of the Torres Strait are as layered as the pearls he cultivates.

He takes us to the cultured pearl beds located around the water of the outer islands in the Torres Straits. Traveling in skinny wooden skiffs with long-shafted outboards mounted at the rear, we head into the Coral Sea to the shores of a deserted island.

The muscular, coal-black pearl divers lower their snorkeling masks, suck in compressed air from a tank and then disappear beneath the boat. They are the free-diving experts of the world.

Several minutes later, they appear at the surface, gasping and holding up a crate of oysters. Kazuie expertly opens and inserts a small semi-circular plastic plug in each oyster. It will irritate the oyster just enough to cover the plug with mother of pearl. After several months, these oysters will be harvested for their meat and half-pearls. Finally, the half-pearls will be glued together to make pearl necklaces, bracelets, and earrings.

It's our third day on the island. We're beginning to feel like we've hopelessly boxed ourselves in and won't be able to get to New Guinea from the island. In a last-ditch effort, Pete goes to the government radio operator, Brian O'Brien, hoping he can help us, and he does.

He tells Pete about a private flight to Daru, New Guinea, in eight days. He also invites us to stay with him and his wife, Kay, until the flight. The O'Brien's guest room is a significant upgrade over our tin hut! We move in immediately, giving the tin shed to our artist friend whose been camping out in his tent on the beach.

The guest room is luxury after the hot, mosquito-infested shed and we return to marital bliss compliments of a closed door and good mattress. This is a brief reprieve from our days of constant traveling. I shop, drink wine, and sunbathe with Kay while Pete enjoys time with the men on the island.

Every day Pete checks at the dock for boats going to New Guinea. However, he'll learn later that the ordinarily busy boat trade from Australia and New Guinea across the Torres Straits stops due to New Guinea's pending independence as a protectorate of Australia. New Guinea will continue to be part of the British Commonwealth, but it will be self-governing. For security reasons and government changes, no one is making the trip.

We'll have to take the expensive flight to Daru and travel from there to Port Moresby. The good news is we'll arrive in time for the independence celebration.

I go to the open-air church where everyone wears brightly colored sarongs and flowers in their hair. The familiar hymns are full of drumming with lots of harmonies but no melody. The proclaimed white teetotaler leading the worship is rumored to be a secret drinker.

Pete goes to Kazuie's island karate class. Kazuie emphasizes using the karate learned in his class only to defend one's self. If anyone uses karate to attack someone, Kazuie'll kick them out of the class immediately and permanently.

I write a story for the Torres News, giving my sanitized view of the island. I leave out "the Thursday Island handshakes". The locals' warmth stood in stark contrast to their brusque handshakes—both shocking and oddly endearing.

Kay and I check out every beach on the island for sunbathing and swimming. Then, Kay and Brian take us to Friday Island to fish using large catch nets. We catch three queen fish and two garfish. They're not good eating but fun to catch. On shore, we find a giant green turtle shell. Too bad we find no live turtles.

Late one night, Pete learns spearfishing from Sampson, a giant Torres Island native. Under the pier lights, he and Sampson spear squid and give them to the women si. Before leaving, Pete insists on buying a red souvenir T-shirt with Thursday Island University written across the front. Beneath the writing, there's a university crest with a picture of a beer stein, a pinup, and a shrimp. It's definitely a classy addition to his wardrobe.

The last night we go to the pub with the two Kens, Captain Webb, the O'Briens, and the Matsumotos. We drink beer and eat boiled shrimp. Ken gives us a lovely pen and ink picture of the island, and the Matsumotos give us a pair of golden-lipped oysters with blister pearls. We've been on the island for two weeks. It will end up being one of the longest stops on our trip and give us some of our fondest people memories.

Early the next morning, we board Kit Thorpe's single-engine plane and leave for Daru, New Guinea, with three auditors and a banker. The half-hour plane ride doesn't seem worth the $66 we pay, but we are at last on the road again.

CHAPTER 10

◈

New Guinea's Colorful Characters

We wing across the Torres Strait. Its turquoise water and green outcroppings shimmer below us. From the northernmost outpost of Australia, Thursday Island, we head to the southernmost reaches of Papua New Guinea, Daru Island. It's two-hundred-seventy miles from Port Moresby, where our cruise ship will leave for Bali in four days. We need to get to Port Moresby quickly.

The plane bumps down on a dirt runway alongside tin shacks and palm trees.

Daru is sadly reminiscent of Thursday Island, except it's abandoned. A lone cassowary standing five feet tall and covered in black hair-like feathers trots through the middle of town tossing its gray helmet-like head atop its long blue neck and wagging its red wattels. We find a sad-looking restaurant and order cokes.

We ask the owner about boats going to Port Moresby. He directs us to Ian Murdock, a trader who travels the nearby rivers and coast buying masks and carvings from the natives. He lives on the outskirts of town. Murdock and his house are larger than most. There's a chicken- wire fence around a small pond. Motors, boats, clay, and wooden masks clutter the yard.

"Halo, mates," rings out as a blond-haired giant emerges from the house. "What are the likes of you doing here in Daru?"

Pete extends his hand to the friendly giant. "We're trying to find a cheap way to Port Moresby."

"You and everyone else in Papua New Guinea! In three days, New Guinea gets its independence from Australia. Even Her Majesty, the Queen of England, is coming to the celebration. I'm staying as far away from Port Moresby as I can." He puts his muscular arms on his hips, looks at us, and continues. "They say all the tribes will come down from the highlands and put on their finery for the sing-sing with the queen. Those tribes aren't civilized. When they get on their warpaint and start drinking the hooch, there's no telling how many spears and arrows will fly." He leans against the fence post scratches his head, and adds. "A freighter's in port headed to Port Moresby later tonight. I know the engineer, Freddie Mathies. He speaks English, and he'll find a spot for you."

"Wow, look at these masks. Where did you get them?" Pete asks.

"I travel into the interior of Guinea in my dugouts, sometimes up the rivers and sometimes among the coastal islands. I trade cloth, beads, and food staples with the tribes. They make the masks for festivals and then throw them away, so they sell them for almost nothing to me."

Ian hands Pete a face mask carved from mahogany. "Take this one. It's chipped. Here's another one for you from the islands. You can tell it's coastal because it has a cassowary breast bone for a face and shells for the eyes and mouth. Look at all the feathers surrounding the face! Can you believe they just throw these away? I can get a pretty penny for these in Port Moresby and even more in Sydney, plus it's quite the adventure to travel inland and negotiate with the natives! Yeh, it can be a bit dangerous, but you see some jungle. I'm looking for some help. Maybe you could ride along with me. I'm leaving tomorrow up the Fly River." Pete and I look wistfully at the large dugout in the canal. The trip tempts us, but it would mean giving up the money we've already spent for our cruise ticket to Bali. The lure of the unknown and doing something no one else might do almost sways us to join Ian.

"Hey, that's an interesting offer, but we've already paid for our passage to Bali. Maybe on our next trip to New Guinea," I answer.

"Things are changing fast here. Get back before it's all gone." Ian warns, "Hey, before you leave, come and meet my pet, Eunice." He takes us to his muddy pond. We wonder, will it be a turtle, a fish, or who knows what kind of jungle animal? He runs a large stick across the chicken wire. Out flies an enormous crocodile. The jaws of this giant, saltwater crocodile push against the flimsy fence. I jump back in fright.

Ian laughs. "She always has this effect on people."

We head to the docks and find a freighter filling up with people. When we ask for Freddie, a deckhand leads us to the pilot house where Jackie Gleason's New Guinea look-a-like greets us. He smiles, and white teeth gleam brightly against his ebony face. His T-shirt has oil stains and a beer logo on it. Bright red shorts cover his stocky legs, and his stubby bare feet hug the deck.

"We need to buy a ticket to Port Moresby. How much will that cost?" Pete asks.

Freddie gives us another hundred-watt smile and waves his hand toward the boat's deck, where people are laying down brightly colored cloth, full baskets, and weapons. "There's only deck passage left."

"How much would that be?" Freddie looks into our young, white faces and then suggests an alternative. "How about you work to pay?"

"Deal!"

"Pete, you can help us load and unload, and Mrs. can cook dinner." We walk up the gangplank and put our backpacks on two bunks in the pilot house.

We spend the night in the pilot house in bunks. As I look back, I wonder about the arrangement. We were too unaware to question his decision. Was Freddie afraid of putting us among natives on the deck? A new era is dawning for New Guinea, but white and black as equals is still a fragile dynamic. Freddie's freighter chugs through the translucent waters, a floating microcosm of New Guinea's contrasts—modern ambition on an ancient stage.

When we arrive in Port Moresby and have nowhere to stay. Freddie invites us to his modern apartment. His girlfriend is a native girl from the highlands with face tattoos. She silently smiles contrasting with our naive exuberance as we buy Freddie his favorite beer, San Miguel, eat barracuda, and laugh—only speaking English. It's a lesson in the unspoken challenges of coexistence.

The next day is New Guinea's Independence Day, and we go out to the street where Queen Elizabeth II and her consort, Prince Philip of England will be passing in procession. It's right next to Freddie's Apartment.

Thousands of chanting, colorfully-clad natives fill the street. We're among them, tightly wedged between the city buildings and the police barricade. Pete and I tower over the shorter natives. We are the only Westerners in the crowd.

New Guinea's Colorful Characters

Prince Philip waves to the large gathering from the back of the Mercedes, resplendent in his military regalia. As he scans the group, he locks eyes on Pete and me with our white faces and bright orange and green backpacks. We are an anomaly among a sea of natives. A year later, we will be in London at the opening of Parliament and see the royal couple again. They'll be in their horse-drawn coach this time, and we'll blend in too well to even get a second glance.

At the football stadium, the New Guinea government has organized a giant sing-sing to celebrate independence and to honor Queen Elizabeth II. A sing-sing is when all the natives from around the country converge on Port Moresby to dance, sing and celebrate. Everyone dresses in their finest feathers and beads.

No royalty has ever been to New Guinea, so there is great excitement. Tribes have been streaming into the capitol all week, and each tribe tries to outdo their enemies and friends with their dancing and outfits.

The legendary mud-men cake themselves in clay and put on their sculpted, large clay heads. Each head is a different face with colorful warpaint and feathers. As we walk past them, they stare, their movements as deliberate and powerful as the history they carry.

A native with a towering headdress of bird of paradise feathers motions me to him. He puts his arm around my shoulders. I make a stiff smile while Pete takes my picture. Other tribesmen in intricately woven cloaks with long trains catch our eye. Women in beaded jewelry and dresses sway to the beating drums. We feel dull, colorless, and underdressed in our Western clothing.

As Freddie's guests we're immediately included in the dancing, singing, and beer drinking. There's nothing like doing the watusi with a painted Moroma tribesman. The sing-sing allows us to see more tribes in New Guinea than a weeklong trip into the highlands. We drink too much and dance wildly. It's a party we won't forget.

Her Majesty misreads the cultural dress code. She wears a pastel, pillbox hat, and matching suit. Where is her ermine robe, crown, and scepter, the natives wonder? They have seen her picture in every government building forever in her finery. Her drab attire is insulting.

The genuine relationships we experienced in this primitive and raw country will be hard to duplicate. Leaving Papua New Guinea behind, I can't shake the feeling that luxury comes at the cost of connection.

CHAPTER 11

Balinese Shangri-La

The joyful sounds of native drumming and European choral music serenade us from the dock as we board the Nieuw Holland. We are leaving the primitive culture of Papua New Guinea behind for the elegance of a European cruise line. I buy a handprinted piece of material in a jungle motif to wear as a sarong for the required formal dinners. I add flowers found daily in my room for jewelry. Unfortunately, my pack has no room for cruise wear, so I make do.

The three-day cruise is a Western-style luxury with warm showers, sheeted beds, gloved service, and five-course meals. Although we've booked the cheapest passage on the lowest deck, where we stay in separate dormitory bunk rooms segregated by gender, we feel spoiled and pampered. We meet Deide and Will, friends from Australia, on the cruise. While they have considerably more resources than we do, we always enjoy spending time with them.

Deide and I arrange it so we're bunkmates. Being together gives us time to catch up, and we talk into the wee hours of the morning.

"We loved your wedding in Bundeena. Especially the music from Bernstein's opera, 'The Mass,' and the Sandpiper song, 'Come Saturday Morning,' which played as you left the church. Who would have thought that little Methodist Church in funky Bundeena would have such an amazing stained-glass window!" Deide says

"I'm glad you could come. You and Will always add to the party. So, what's happening between you and Will?" I ask.

"Neither of us are ready to commit, but we're enjoying traveling together. I just have to say, you looked gorgeous in your cream-colored dress, but ohhh, that picture hat was magnificent with the chocolate silk sash and flowers."

"Thanks, I spent more on the hat than on the dress, but you know how much I love hats."

"That gold cross you were wearing was perfect with the dress. Where did you get it? I've never seen it before."

"Pete's mom gave it to me. The origins of the cross go back to their Danish heritage. Queen Dagmar, a beloved 12th-century Danish queen, was buried wearing the cross. She bravely wore it to proclaim her Christianity in a country still heavily Pagan."

"Were there any minor crises with the wedding?"

"Let's see, my new in-laws forgot about the international date line and appeared at our door in the morning a day after we expected them, which blew our cover. They thought I was living with my friends and Pete was living with his. We'd been living together all year.

We took Steen and Mary Anne to the beach in Bundeena for a swim. The week before they came to Australia their kids gave them a 25th wedding party where they gave their mom a ring with the birthstones of all seven of her kids. While we were swimming, she lost the ring. We looked for it, but it was hopeless on the sandy beach.

Obviously, the ring must have meant a lot to her, but she said, "It's just a thing. My kids showing their love for me is more important. They'll understand." Steen hugged his wife and said he'd try to replace the ring.

"Wow, it sounds like your in-laws are great."

"They are. How did you and Will get to Port Moresby?"

"We caught a plane from Cairns to Port Moresby and then took a tour of the New Guinea Highlands."

"Sounds expensive!"

"It was, but you know, Will has his trust fund he can always dip into. We flew three hours into the interior in a small Cessna with a pilot and guide and stayed at a lodge that blended into the mountains with lovely woodwork and modern facilities. We took bird-watching hikes and saw lorikeets, birds of paradise, kookaburras, and magpies. There were huge native pines that towered over the canopy, giant banana trees, and huge ferns. Even though we were in the mountains, it was hot and humid. My favorite part was visiting the native village. Tomorrow, we met the chief and were able to dance with the locals. They wore fabulous outfits with bones, fur, feathers, and grass skirts."

"I bet you boogied with the best of them, friend."

At meals, all four of us share where we've been and what has happened since our last dinner in Cairns. Whose travels are more adventurous? We try to out-adventure and story-tell each other. It's good to see some familiar faces again.

Also traveling steerage are Zee and LeAnn. He's a gray-haired hippie, and LeAnn is a free-thinking teacher from San Francisco. They had also been working in Australia and were returning home to the us. They join us. We will have deep conversations about Ayn Rand, Transcendental Meditation, LSD, and the Vietnam War. The luxury of the cruise and our jam sessions with fellow travelers is a hiatus and reminder of the America we left. Tomorrow we dock in Bali.

The laughter of the New Guinea tribesmen echoes in my memory—a sharp contrast with the understated politeness on the cruise. The warmth Ian, Freddie, his girlfriend and the tribesmen showed us felt genuine and real next to the polished coolness of our fellow passengers.

In the morning, we wake up in the Shangri-la of the South Pacific, Bali. From our cruise ship, we see a white Hindu temple standing out against the lush green of the shoreline. It's a layered wedding cake decorated with intricate frosting. As our tender nears the private dock, the Balinese band plays exotic and atonal music. Gongs, a xylophone, tambourines, bongo drums, and flutes provide the beat for six lovely Balinese dancers in green and gold-threaded sarongs who place leis of jasmine flowers around our necks as we step out of the tender onto the dock.

Once through the gated cruise entrance to the dock, chaos breaks loose as taxis, bemos, minibuses, tourist buses, and private cars vie for passengers, using their horns and megaphones. We pass the air-conditioned tourist buses and climb into a bemo, a skinny truck with a canvas roof over two passenger benches facing each other. We barter for a price but know we'll pay more as incoming tourists.

Bypassing the closest city, Denpasar, we head for Kuta Beach. In 1974 there was only one high-rise resort along the beach and many losmens, small guest houses, each uniquely built with carved teak and mahogany. There's always a fountain at the entrance surrounded by orchids and bamboo.

Balinese Shangri-La

We hear a chorus of yahoos and Rhonda Murdoch's distinctive dog whistle as we pass six bikini-clad girls with sarongs, sunglasses, and large sun hats. Finally, by pounding on the side of the bemo, we get the driver to stop. Our teacher friends from Sydney have made it from Darwin to Portuguese Timor and Bali!

"'Ow are yah, mates? Welcome to Paradise!" Rhonda yells as she reaches out to us with a hug, holding her cigarette carefully to the side.

"Welcome, welcome, welcome!" Haley adds. "You have to stay at our guest house. It's near here, and it's so perfect. It's on the beach with free breakfast. A young couple runs the place." She pulls on my arm, and we head toward the guest house while the other four girls continue to the beach.

Pete rushes ahead in the direction Haley points.

"Hey, thanks for forwarding our mail from Darwin to Port Moresby, Haley! It was so good to read the letters from home. How was your trip to Darwin and Portuguese Timor?"

Haley answers, "Rough, I wished for the mini moke trip again, if you can believe it. We took the short and cheap flight from Darwin to Dili in Portuguese Timor. From there, we found a small cargo ship to take us to Bali. Five single women dressed in Australian beachwear, not speaking the language, on a boat with a filthy hole in the floor for a bathroom, and leering men. Kevin's chauvinistic rantings in Australia were nothing compared to the grabbing hands and sexual innuendos of the crew on our cheap passage! I complained about Kevin as a travel companion, but try putting up with whining, spoiled women who've never traveled anywhere outside of Australia. Helen, Kathy, and Carmen can't stop comparing everything here to the comforts of Australia."

"Bali is easy. I'm unsure how well they'll do once we get to Djakarta. Rhonda is tough, and I'm counting on her to be with me the whole trip," Haley adds.

I updated her with my news. "Pete and I are still married, but I'm pressing to make the trip more mine, not just his. He likes to keep going. I like staying in places and hiking or taking a course. He seemed surprised to find out we like different things."

Kuta Beach is a youth scene with hippies and lots of partying. Haley and her Australian girlfriends are in the middle of the action. Bali is the holy grail for overland travelers starting in Europe and going through Asia. After a lot of hard travel, it's Nirvana with good food, friendly natives, an endless beach, and lots of dope and other drugs.

Our first night in Bali is a festive reunion with our Australian teacher friends and our Canadian friends, Deide and Will. We start with a beach sunset and move to Poppies, the newest restaurant choice for Westerners. We order Gado Gado, a local vegetarian salad with a rich peanut sauce. It will become Pete's favorite meal choice in Indonesia. The meal costs us three hundred rupiahs which in US currency is seventy-five cents. Losmens are two hundred rupiahs or fifty cents a night. "Giggle grass," as they call marijuana, is two-thousand Indonesian rupiahs for an ounce, about five US dollars. Pete buys Haley's Portuguese Timorese Escudos. His travel goal is to collect each country's currency and ancient coins. In 1975 the Escudo would no longer exist when the Indonesians invaded East Timor and took it over. Pete's purchase becomes unusable but rare.

It's no wonder there are so many young ex-pats from all over the world living the hippie life in Bali. The hipsters smoke dope, lounge, go to the beach, and eat. We can handle only so many sunsets with tropical drinks in hand. The next day we rent a motorcycle to explore the rest of the island.

We head to the island's interior and the mountains early in the morning. We pass through small towns, each of which has its own particular craft; the stone carvers, weavers, gold and silver jewelers, and painters. We recognize their world-class craftmanship, but hesitate on spending too much money so early in our trip.

Above the Tampansiring temple are springs and Sukarno's summer home. Sukarno was the first president of Indonesia when it gained independence from the Dutch.

The temple is open and smells of sandalwood and jasmine. We meet a young Balinese man studying Balinese music at the University of California in Los Angeles. He attempts to explain the religion of Bali, which is a blend of Hinduism and Animism. Every morning in Balinese households, rice and flowers are offered at small shrines to appease the evil anima and attract the good anima. Spirits live in rocks, statues, and trees. To avoid being disturbed by these pneumas, they wrap poleng cloth (black-and-white checkered gauze) around the natural objects. When praying, the Balinese seem to be in a deep trance.

The road is paved and uncrowded, and the day is warm and sunny. As we climb up the mountainside, the temperature drops, and we stop for the night in Kintamani. The guesthouse sits on the lip of a volcanic crater with a lake below.

In the morning, we hire a guide who speaks English to take us to the native village across the lake. He paddles a dugout canoe to the small settlement which sits on a narrow strip of land between the lake and the crater. Their temple is in a banyan tree. Several thatched-roofed huts hug the rock wall encircling a small dirt square. At the end of the spit of land are high platforms where they put their dead. They are a unique tribe separate from mainstream Balinese Hindus who cremate their dead. Farther along the shore, we stop at a hot spring for a dip and picnic lunch. Finally, we go back to the guesthouse, pick up our packs, and hop on our motorcycle for Singaraja on the opposite side of the island from Denpasar.

The sky darkens, and monsoon rain pours down on us! We quickly hop off the motorcycle and head for a shed which we share with three calm water buffalos. After ten minutes, the sun reappears, we continue our journey.

In Singaraja, we go to the market where Pete finds an old woman surrounded by large wicker baskets full of hundred-year-old Chinese coins called "qian" or "cash." They are round copper coins with square centers which the Balinese toss like confetti at funerals. For just a few dollars, you can buy a kilo of them. Pete picks through the baskets looking for ones in good condition. His search makes the seller nervous because no one has inspected her coins so carefully before. Pete quickly adds his handful of coins to a kilo bag, buys them without haggling, and we leave.

That night we celebrate my birthday by splurging on an expensive Chinese restaurant meal, but the menu gives us no idea what to order, so we watch plates coming from the kitchen and point to get what we want.

Pete pulls a package from his pack. "Happy birthday, Miss Rohver. I didn't forget." Underneath the tissue paper is a praying man carved from a lustrous chunk of mahogany. I hold it up to the lamp light.

"How did you buy this without me knowing? It's beautiful, Pete. I love it. Thank you. It's perfect."

As we walk back to our guesthouse, the air is soft with a rich, spicy smell. We hold hands and feel wonder and delight at being in this place.

We have one last thing to do in Bali before we leave; go to the Sangeh Monkey Temple just outside Denpasar. It's a long thirty-minute bemo ride from Kuta Beach. Sangeh is a large forest preserve where gray long-tail macaques monkeys guard the Bukit Dari Temple.

Three young Balinese men are standing in the forest's parking lot. They're dressed in traditional Balinese dress, sandals, gold-hemmed sarongs, collared short-sleeved shirts, and gold-threaded red turbans with material tufts in front. In their hands, they hold bags of peanuts to feed the monkeys. I give the vendor several rupees, and he passes me two bags of nuts. Monkeys fly at me from all directions as the bags touch my hands! I shriek and throw the peanuts in the air. It's a terror filled moment for me.

The peanut vendors smirk, full well knowing what the monkeys would do.

In Kuta Beach, we say cheerio to Haley, Rhonda, Helen, Kathy, and Carmen. Maybe, our paths will cross again. Already Kathy and Carmen are talking about flying back to Australia to see their boyfriends. I hope Haley won't end up traveling alone.

Tomorrow morning we're traveling to Jogjakarta the country's cultural center on the Indonesian Island of Java. Jogjakarta is the only Indonesian city where a hereditary monarch still rules. Jogjakarta was the Dutch, colonial name for the city in 1974. Today it's called Yogyakarta, a name from its historical past.

I plan to take a week-long batik class in the city while Pete looks for coins and explores the area. We wonder if Java will have the same magical feel as Bali or its own unique magic.

CHAPTER 12

Batikking in Jogjakarta

A van takes us from our hostel in Kuta to Gilimanuk, where the ferry leaves for Ketapang on the island of Java. The landscape changes as we head toward the Bali Strait. Stone altars and well-tended gardens give way to wooden gates and weeds. We wait for the ferry, and while waiting, I dunk in the crystal-clear waters of the Strait, where colorful tropical fish swim around me. Pete collects coins. I keep track of all the exotic bodies of water where I swim.

The island of Java is only 2.4 kilometers from Bali; a quick trip. The strait connects the Indian Ocean with the Bali Sea. From the dock to the train station is an easy walk. The train is old with wicker seats and plenty of leg room.

We're facing a row of young children who are making funny faces at each other. They squeal with laughter when we make finny faces, too. No language, but we're communicating.

Outside the train window blocks of lime-green rice paddies neatly separated by berms of darker green fly by. It's dry season; planting time so there are workers in the fields men with their water buffalos plow through the muddy soil while groups of villagers follow doing the backbreaking work of planting the seedlings one-by-one.

The dramatic Javanese landscape glides by with its dark volcanic mountains and green hills. Mt Ijen volcano, the "lonely crater", spews white steam into the sky.

A palm-lined road crosses the tracks. People wait on both sides for the train to pass. Bare-chested men wearing turbans and colorful sarongs stand astride battered bikes. Women balancing bunches of bananas on their heads with children strapped to their backs look curiously at us as we speed by.

"I'm getting excited about learning how to do batik," I say to Pete as we share a packet of rice and dried vegetables. "I don't know anything about batik material or pictures except I like the way they look."

"You pick up things quickly. You'll do fine. I'm counting on your batik class giving you first-hand knowledge on what to look for in a quality batik painting. Maybe your batik will be so good we won't need to buy one," he encouragingly says.

"Thanks for the vote of confidence," I say smiling at him. "I don't think I'll be a batik master after a week of instruction, but I certainly can find out the difference between a one-dollar Batik picture and a twenty-dollar one. It feels good to stop in one place for a while and learn something new. We've been going fast lately; looking, taking pictures, and moving on."

"What! You make it sound touristy. We're not just checking off destination boxes. We're meeting interesting people, learning new words, smelling new smells, eating new foods and experiencing things we'd never be able to do anywhere else."

"Yeh, you're right, but I'm happy to be staying in Jogjakarta for a week." As we continue to plan the trip together, I feel listened to and considered; our relationship easily stretches between us.

Batikking in Jogjakarta

The trip from Denpasar to Jogjakarta takes eight hours. In one day, we go from mini-bus to ferry to train.

Our new hostel is a perfect mix of comfort and local charm. Although we're tired after our trip, we decide to walk to the Batik Research Center and sign me up for a batik course.

The center has been reviving the art of Indonesian batik and offers short courses to tourists. Jogjakarta is the Javanese center for fine arts and culture. It also has the best university in the country. Students, international backpackers, artists, and musicians, are drawn here to learn ballet, batiking, textiles, drama, literature, music, poetry, silversmithing, visual art, and wayang (shadow) puppetry.

I meet Bambong Darmo, a batik artist, and sign up for his weeklong course. Dianne is another one of his students on a semester abroad with Sarah Lawrence College in NYC. She helps both Bambong and me to communicate. At first, I find her cynical and aloof, but she becomes my real batik instructor and a good friend.

Pete and I eat at a Chinese restaurant with new Swiss friends, Gisela and Roger, that night. The Minders are a couple from the German sector of Switzerland. Roger builds cameras, and Gisela is a textile designer. Pete will spend the week going around Jogjakarta with them.

We watch several frog legs hopping in the large woks but stick to more conventional fare. After, we meet up with Haley's traveling women's group. They're headed down through Java to Djakarta. They only stay in Jogjakarta for a day. We will not see them again in our travels.

The group is down from six to four women. Haley, Peg, Rhonda, and Helen are still traveling. Two have flown back to Australia from Bali to be with their boyfriends. For them, the adventure is over after just traveling through Portuguese Timor and Bali, Indonesia. We go together to a US movie with Indonesian subtitles. We laugh, and then we hear an echoing laugh from the Indonesians. They must catch up with what's happening by reading the Indonesian subtitles.

The next day I go to Bambong's studio. It's in a garden behind his house. There's a large, covered verandah with a dais at one end and several fire pits heating huge copper caldrons bubbling off to one side. Wet, newly-dyed batik cloth flaps behind the caldrons.

Bambong hands me a chanting (a copper funnel with a bamboo handle). "This canting. Draw with wax on cotton," he says leading me to a small stool and an easel. He waves his hand toward the craftswomen in the courtyard.

Bambong's sweeping gesture toward the batikkers leaves me floundering. The women's laughter rings out, a gentle reminder that sometimes you learn best by doing. I dip the chanting in the wax and begin.

The wax drips stubbornly mocking my attempts at precision. Still, I keep going, buoyed by Diana's encouragement and the rhythmic hum of the studio.

Diana is my batik instructor. She shows me how to hold the canting and takes me on a studio tour, explaining the batik process as she goes. "First, apply wax in the places you don't want dye. You begin with the darkest color. Put the cloth in the dye vat until it's the color you like, dry it, and then boil off the wax. The fabric will be white where the wax was. You can repeat the process many times with a multitude of colors until you have the design you want."

Traditional Indonesian batik uses white, brown, blue, and black. There are also traditional patterns. I make a checkerboard and fill each square with a classic design. I admire the skill of the Javanese women as they bring to life the designs Bambong draws. I learn to sweep and swoop expertly with the chanting, but I don't even come close to the detailed work of my fellow batikkers.

While I'm heading off every morning for my batik class, Pete, Roger, and Gisela explore the city together. It's clean, engaging, and well laid out. While Pete scavenges the market for old coins, Roger uses his expensive camera and different lenses to capture images of the people and place. He gives Pete photo tips. Later in Singapore, he will help us buy the best camera we've ever owned. Gisela buys batik cloth for her textile collection.

They visit Kraton, the sultan's palace, where the royal family still lives in the center of the old city. The sultan's palace gleams in the sunlight, its open courtyards alive with gamelan music and the scent of frangipani. Ancient artifacts and historic batik pictures decorate the courtyard. The artifacts, supposedly, ward off evil reminding us of Bali's animism.

That night, we eat Gado Gado, a large green salad with bean sprouts, lettuce, and eggs, smothered by a spicy peanut sauce, and satay, skewered meat grilled with spices. Inside the restaurant, a gamelan percussion orchestra plays. Gongs that look like bronze pots with lids chime. Met allophones, a xylophone with metal rather than wooden keys, tinkle. Two-stringed fiddles plunk. The drums thump. A single flute trills. The instruments' exotic sounds fill the dark, humid night.

One day the Minders and Pete go out to the 7th-century ruins of enchanting Borobudur. It is the world's largest Buddhist temple, with Hindu and Buddhist elements. The stupas and mandalas are Buddhist, and the temples are Hindu. The temple complex lay buried under volcanic ash for eight-hundred years until it was rediscovered in 1911 by the British governor, Lieutenant Thomas Raffles. In 1974, the ruins are neglected and unknown outside of Indonesia. Today they're a major tourist attraction.

The night before we leave, a student from the ship, Semester at Sea stays in our hostel. He takes the train from Jakarta where the ship is docked. Semester at Sea is an American college program where a college student can earn credits by traveling to different places and taking classes onboard.

Ben, our student friend, looks like a British explorer from the 1900s dressed in cargo pants, wide straw hat, and hiking boots. The twenty-year-old is infectiously curious and enthusiastic, like a puppy venturing out of his den for the first time. Jogjakarta is his first trip alone away from the ship without a guide. Once he arrives at the hostel, the bathroom is his first stop.

"Hey, where's the toilet?" he yells through the door.

"See the hole in the floor with two rectangles on either side? Place your feet on the rectangles, lower your butt, and aim. The water jug on your right is toilet paper," Peter yells through the closed door.

A short time later, we hear splashing and yelling, "How do I get out of this bathtub?" Rushing through the door we find the young Nebraska boy lathered up in the middle of the cistern, grasping the brick walls in a futile attempt to climb out.

"It's not a bathtub. It's the water supply for the hostel, and you've just dirtied it up," Pete pulls him out of the cistern. "Use the scoop on the wall and throw water over yourself to get the soap off. I'll go get the hostel owner so he can clean up the cistern tank."

"Gee, I'm sorry. Indonesian bathrooms are strange."

"Yes. They're different from ours. That's for sure!" Some cultural lessons are more embarrassing than others.

The next night, we take the overnight train to Djakarta with our young friend from Omaha. As a reward for Pete's bathroom lesson, he invites us for lunch on the ship docked in Djakarta. We might welcome refuge in this huge, boisterous city. I've learned a new skill. I fill my pack with batik materials and good intentions, which I will never use again.

CHAPTER 13

Singapore Slings at Raffles and Ribs at the Petroleum Club

Pete and I are next in line for Singapore's immigration. As we walk to it together, a guard approaches me, a rifle slung over his shoulder, holding up one hand and pointing at the line. "Only one at a time," he says, motioning me back behind the black tape. After Pete is through, I go up to the counter.

She stamps my passport, writes in some numbers, and says flatly, "Have a good visit."

I stare at the stamp—three days? I can hardly believe it. "How long is your visa?" I ask Pete.

"Two weeks."

"Why do you think they gave me only three days?"

My indignation grows as Pete says, "They must like me better."

We go back to the immigration booth. Pete insists we go together to the counter this time. He hands the officer both our passports and asks, "Why did I get two weeks and my wife only got three days?"

"Not same name on passports." The officer's apology is half-hearted: Many single women come here for … business. He wags his head back and forth and hands back our passports. It's a reminder that in some places, being an independent woman comes with suspicion.

"Guess they don't think you're wife material," Pete jokes.

"They obviously think I'm hot enough to do business here," I reply.

"I can see you're getting hotter, so let's move on. We'll only be staying two days in Singapore and then taking a car trip with the Minders to Malaysia, so your three-day visa won't be a problem."

"You're right, but maybe it's time I became your wife on paper as well as in bed."

We're met at the airport by Kay Hawkins and her driver in her air-conditioned Cadillac. The Hawkins are friends of one of my student's parents in Australia who came to our Bundeena wedding and invited us to stay with them in Singapore. Art supplies oil field equipment for the local oil business. The couple call themselves Texas "oil field trash," but their lavish ex-pat life is anything but trashy. They live in a large house in the wealthy Changi neighborhood. Kay is excited about showing us Singapore.

Singapore dazzles us with its contrasts—modern skyscrapers jostling against colonial relics and vibrant temples amidst bustling markets. Long Bamboo poles draped with laundry stick out the windows of tall apartment buildings like welcoming flags. Racecourse Road, climbs up to the top of Mt. Faber giving us a spectacular view of the city. En route we stop at the temple garden. The Temple of lights glows with a serene majesty, its golden Buddha radiating peace. It is a quiet contrast to the bustling streets of Chinatown. Kay then treats us to a lunch of curried crab, a Singapore specialty.

We stop at Changi Prison, a Japanese internment camp from WW ll. Walking through prison, the weight of history presses on me. The past is alive here, a reminder of resilience and survival.

Before going back to her house, we stop at the American Embassy and they quickly change my passport into my married name. I'm now officially Mrs. Elizabeth Kristina Fischer. That night they take us for dinner to the elegant Petroleum Club where we eat prime rib and rub elbows with oil executives and diplomats.

The next day, we leave on a seven-day road trip along the east coast of Malaysia. (see next chapter 14)

When we return, the Hawkins, in true Western-style, welcome us and our Swiss friends, Gisela and Roger Minder, with barbecued steaks, corn-on-the-cob, and refried beans. The following day the Minders fly out of Singapore to Los Angeles, where they'll stay with Pete's family in Calabasas. We've made plans to meet up again six months later at their home in Dotzigen, Switzerland.

While in Singapore, we send several packages to the US. Our packs are getting too full. They wrap our boxes in white cloth at the post office and sew them shut. On our return home, nothing will be missing.

We go to the Chinese Products Emporium, a shopping mall in an old hotel building in Singapore's China Town. Pete, amazingly, finds a 1,000-page hardcover book about world coins. So much for lightening our packs!

We're anxious to see how our slides we've taken with our new camera have turned out. Aside from a few accidental hand and foot shots, they're good. We'll send the slides home to our families in our packages.

Kay insists we visit the Tiger Balm Gardens, a large theme park with 1,000 statues and a series of dioramas depicting aspects of Confucianism and Chinese mythology. Plastic, brightly colored dragons, pagodas, tigers, Buddhas, and fighting men fill cavern-like exhibits. The maker of Tiger Balm is responsible for this commercial park. Tiger balm is a waxy mixture of pain-relieving herbs in a tiny round tin found everywhere in Asia and the US. In 2004 they tear down the garish gardens for redevelopment; a change for the better.

No trip to Singapore is complete without a trip to the Long Bar at the Raffles Hotel. It has a plantation feel with bamboo furniture once made at Changi Prison. The mahogany bar stretches the length of the room. Above it, slow-moving fans turn. A wide veranda overlooks the tropical garden.

Today the bar at Raffles is much the same way it was in 1974. The hotel still takes up an entire city block of prime Singapore real estate. It's been designated a World Heritage site and National Monument of Singapore. Raffles is one of the few 19th-century hotels left in the world. In 2018, renovations started to keep its 19th-century features but to also modernize it for the 21st century.

To thank Kay, we buy her a Singapore Sling, the drink invented here by the bartender Ngian Tong Boon in 1915. The drink has Gin, four different liquors, lime, and pineapple juice which is shaken and served in a highball glass decorated with a maraschino cherry, a chunk of pineapple, and a paper umbrella. It was initially considered a ladies' drink because of its pink color and clear liquor, but now everyone has one when visiting the bar. We toast Kay and others who have frequented the bar.

"Here's to Frank Buck, who made Raffles his headquarters in the nineteen thirties when he wasn't out hunting and collecting exotic animals for zoos worldwide!" Pete adds.

The bartender joins the toasting, "Here, here to Queen Elizabeth II, Liz Taylor, and Michael Jackson, more recent guests!" Being here is like stepping back into history. We savor the moment.

Kay stands, looks toward the lobby, and eloquently honors a literary great from the past. "Here's to W. Somerset Maugham, who said, 'Raffles stands for all the fables of the exotic East.'"

It's our last night with the Hawkins. Kay and Art's generosity leave a lasting legacy of hospitality for us to replicate in the future. We go out for dinner at a barbecue restaurant with a Country Western Chinese band. Stomping in a line dance contrasts sharply with our surroundings. Emigrants always bring pieces of home with them We'll take the train to Penang, Malaysia in the morning.

"Kris, Art offered me a job tonight," Pete tells me when we're alone in our room. "What do you think of that? I'd be working for Smith Tool Company, the world's biggest supplier of drilling bits!"

"Great, we'd be staying here in Singapore, right?"

"I'm not sure, but probably."

"Well, do you want to? Remember, your dad wanted you to return to the US after our wedding. He said two years was plenty long enough to be away from home. He wanted you to start a career. Get work experience. Set a good example for your six younger brothers and sisters. Remember?"

"This would be work experience! You could get a teaching job somewhere, but we still haven't seen India, Nepal, Afghanistan, Burma, or Thailand. We've just started our trip."

"We could see those places later."

"Later, that's what I'm afraid of. Right now we have the freedom and time before we have mortgages, careers, and kids."

"If you think you'll regret not finishing our trip, don't take the job. You rarely get a rewind in life."

"I would resent staying here. Traveling around the world has always been my dream and I'm doing it! It's not time to stop. I'll let Art know tomorrow morning."

CHAPTER 14

Breathtaking Beaches, Jungle, Pirates, and Barbed Wire Barricades

After a week in densely populated Jakarta and Singapore, we're ready to explore outside the cities. We rent a car with our friends, the Minders, to drive down the east coast of Malaysia which is known for its unspoiled beaches and fishing villages.

The Minders are older than us by ten years. We look up to them because they have established careers, speak several languages, and have thoroughly planned and researched this trip. Roger has helped us buy a Minolta single lens reflex camera in Singapore, and has promised to teach us how to use it during the trip. They know so much and are in sharp contrast with other travelers who are floating through their trips with no plans or goals.

Singapore traffic tangles us up, slowing down the start of our trip. But once beyond the city limits, the Malaysian countryside opens up shimmering with emerald fields and terraced hills; the light playing tricks on our eyes as it dances across the landscape.

After four hours, we reach Mersing. In 1974, it's a small kampong, or village, with fishermen carrying nets to their rainbow painted boats lining the beach. Roger runs to the beach his camera blazing. "What a place; a thousand shots!" he yells to us. The boats' prows stick upwards, each with a tethered kite attached to it as a wind telltale. Waves splash over the square sterns. The fishermen wearing batik shirts and sarongs, prepare to go out with the tide change. The sarong also known here as a sapin is rolled and tied in an intricate, ceremonial knot around their waists. Even with hard physical work, the sarong stays tightly fastened.

Bamboo fish traps are stacked on the beach drying. A man salts fish. A group of women are discreetly bathing behind six-foot-long saris held at each end by two other women. Several children sit naked on a log, giggling and pointing at us. We give them some hard candy and they giggle even more. An overwhelming feeling of awe and wonder comes over me as I stand in this exotic and faraway place; feeling both free and fortunate.

We stop at a thatched stand and sit on palm tree stumps as we eat fresh melon and drink coconut water. A small boy shimmies up a tree and drops a coconut at our feet. We reward him with candy and hand the coconut to the woman at the stall.

The people of the fishing villages are mostly Austronesian Malays who remind me of our Hawaiian Polynesians. The woman who serves us wears a tropical-flowered sarong tied above her chest.

As Roger fiddles with the new Minolta, Gisela teases him mercilessly, "A thousand shots and not one worth framing!" Our laughter mingles with the sound of crashing waves. Palm trees edge the white sand beach, shading the traditional Malaysian houses behind them. They are called "attaphouses" because their thatched roofs are made of attap leaves from a flowering palm tree found along the ocean coast. The traditional Malay house sits on stilts, sided with hard wooden overlapping planks, lighted by kerosene lamps, and with an outhouse in the back. Initially, Malays build the main section of the house called "rumor ibue" or "mother house" and later make extensions to it, such as a guest porch and kitchen. We'll be staying in a tin-roofed guest house with electricity and bathrooms. It will have the distinct Malay house design with a pointed peak swooping down to a winged roof; and a tiled front door with an entrance staircase that reaches out like welcoming arms.

In the restaurant for dinner, we can't read the menu in Malay, once again, we spy on the plates being brought out to other customers and point to what we want. Most of our dishes consist of a glutinous rice with fish in curry sauce, or fried.

The next day we travel five hours to Batu Rakit, known for its artisans who make songkets (felt pillbox men's hats), batik cloth, brass lamps, wood carvings, woven mats and boats.

We travel along the modern and well-maintained coast road going fast until we come to rivers where small three-car ferries take us slowly across. The boatmen pull the ferries with gloved hands gripping heavy ropes stretched across the swift, murky rivers. The roads contrast sharply with the ferries: modern clashing with past.

It's a good break from driving. We soak up the hot moist air blowing across the barge, look at the trees of the jungle reaching over the river, and smell the earthy, green growth.

When we get to Batu Rakit our first stop is a batik factory. Malaysian batik is made from handheld wooden blocks carved with tropical flowers, leaves, and butterfly designs. The blocks are dipped in wax, then stamped on thin cotton material which is dyed or painted.

To create multi-colored or more complex patterns, different wood blocks are used with several re-dyings and paintings. Huge sheets of fabric printed with red, purple, yellow, and blue flowers flap in the breeze around the pavilion where the women sit beneath an open thatch-roofed shelter, transforming the white cloth into colorful gardens with their dyes and blocks. Gisela, who works in the fashion industry in Bern, buys yards of this exotic material.

"Do you think you have enough colored material to transform Bern's black/beige style?" I ask Gisela.

"This is a good start. The West needs to wake up their palette and this is a beginning," she answers.

We then go to the boat builders where without blueprints or nails, the men make not only local fishing boats, but also ocean-going yachts sought after by sailors around the world. The builders' reputation for craftsmanship and the hard cengal wood is appreciated worldwide. Roger is amazed by their work using only hand tools. It's hard to impress this perfectionist Swiss man, but the Malay workmen have.

On the beach we enjoy another fish meal and watch an interesting volleyball game, called Sepak Takraw played with hands and feet. We buy some colorful, thin mats to sleep on, and stay at another guest house on the beach.

The next day, we head inland at Kuantan to Malaysia's largest and oldest national park, Taman Negara. It's the oldest rain forest in the world, having been in existence for one-hundred-thirty-million years, where dinosaurs roamed once. To say it is diverse is an understatement. In the park you can find fifteen thousand plant species and six hundred and seventy-five kinds of birds, not to mention mouse deer, barking deer, Malayan tapirs, marque monkeys, butterflies, wild boars, orangutans, clouded leopards, Borneo pygmy elephants, Bornean rhinoceros, and the Malaysian national animal, the Malayan tiger. We stay in a guest house in Jerantut, just outside the park. At night, sounds of the jungle surround us. Was that an owl? Is that crashing sound a tapir or a gaur, Malaysia's wild cattle? Is that scream from a frightened marque monkey? What is stalking it from above or below; a reticulated python or a Sumatran spitting cobra?

It's noisy all night, but the next morning when we take a short walk, we hear almost nothing. A fluorescent blue butterfly with wings trimmed in black with red spots lands on my arm. Our new camera captures the moment. There's a cackling sound and a magnificent Malayan peacock pheasant with large emerald-colored spots on its tail crosses the path. We have left too little time to explore this jungle and its creatures.

We head for Kuala Lumpur. Along the way, we're stopped several times at barbed wire barricades across the road by soldiers with machine guns. It makes us wonder how safe traveling by car is in Malaysia. Two months later in June of 1974, two top police brass will be ambushed and killed by Communist guerrillas in this area.

Although Kuala Lumpur is the capital of Malaysia, it's relatively young and small. In 1974, its only been a city for a hundred years, and its population is around six hundred thousand. Its first inhabitants were eighty-seven Chinese tin workers who arrived in 1857. Today it's an Asian powerhouse, boasting seven-million people and the Petronas Towers, which are among the tallest buildings in the world.

We drive through the central district along almost identical, colonial government buildings and the main mosque. They were built by two British architects named Arthur Benison Hubback and Arthur

Norman. They both combined gothic, western, and Moorish styles in the buildings, with alternating layers of red bricks and then cream-colored tile. All the buildings, the Federated Malay State Railways building, Jamek Mosque, the General Post Office, and the Sultan Abdul Samad Building have graceful arches, curved colonnades topped with shiny copper cupolas, and grand entrances.

The Sultan Abdul Samad Building also boasts a forty-one-meter high (135 feet) clock tower modeled after Big Ben in London. In 1999, the government moved to Putrajaya, fifteen miles south of KL because the city's population had grown to almost four million. In 2022, all or most of the government buildings that we admired in 1974 are no longer used by the government. Some of them have been turned into museums and others are empty.

We cross the Kelang River and follow the green belt that stretches through the city called Lake Gardens. We admire the National Museum, that has a tile depiction over its entrance of its history, from the Stone Age through the ancient Hindu/Buddhist states, to the Muslim Sultanate of Malacca, and the British Colonial Period, and finally to its independence as a constitutional monarchy in 1957.

We stop to admire the national monument honoring the soldiers who died in World War I and II, and those who died in the Malaysian or Communist Emergency. We all wonder what the Malaysian Emergency is. I will later read about the Emergency in the book, War of the Running Dogs. Later in Istanbul, I will meet a young Malaysian who will tell me more about the time from 1948 to 1960 when Malaysia, like many Southeast Asian countries, was being overrun by Communist guerrillas.

Pete suggests, "Let's stop at these stalls and eat some lunch."

"Do you think that's safe?" Gisela replies.

"The stalls and people are clean, and it smells delicious," Pete says.

"We were warned not to eat anything off the street!" Roger protests.

"I'm in," I say as I head to a stall with a smoking grill.

"I know you're careful about these things, but actually I think this is faster and smarter than eating in a restaurant. Have you looked in any of those restaurant kitchens we've eaten in? At least here we get to see them making it in front of us," Pete says, and then follows me.

"All right, let's try it, Gisela," Roger says reluctantly.

We start with satay which is grilled seasoned meat on a bamboo stick with sweet and spicy peanut sauce. Then we try a bowl of nasi lemak: coconut milk rice, boiled egg, anchovies, roasted peanuts topped with sweet and spicy sambal paste. For dessert we buy the stinky durian fruit and have the vendor take off its spiny flesh, leaving us pieces of the smooth, soft fruit with its heavenly, custard flavor. We're told by the fruit vendor. "Durians smell like hell, but taste like heaven."

"Mmmm, that was delicious! Hot, spicy, fresh and quick," Gisela announces. We lick our fingers, finish our bottles of water, and hop in the car. Pete and I both look at each other. We're hoping that our friends won't regret this jump into Malaysian street food later tonight.

We're on our way to Malacca where we'll spend the night before heading back to Singapore. Our time in Kuala Lumpur has been rushed, but we come away liking this clean, charming city with its wide streets, low buildings, and green parks.

Luang Prabang, Marrakesh, Rawalpindi and Malacca are places that conjure up images of exotic intrigue, and we're now headed to

Malacca. Here I am, a small-town girl who spent eighteen years of her life in a rural town in western New York, seeing one of the fabled cities of the world. I pinch myself.

Malacca was once one of the richest ports of the world connecting West to East and a part of the maritime Silk Road. Malacca is a cosmopolitan melange of many cultures dating back to its founding. Chinese, Arab, Indian, Persian, Turkish, Portuguese, Dutch, Siamese, and Malays have occupied the same space for centuries. As the locals say they're "mixed up in a good way."

We enter Malacca. It's located on the sluggish Melaka River where it goes into the Malacca Strait. The Malacca Strait is a narrow strip of water five hundred and eighty miles in length strategically located between the Malay Peninsula and the Indonesian island of Sumatra. It's the shortest and main shipping channel between the Indian and Pacific Oceans, and one of the most important shipping lanes in the world. The Malaccan pirates are notorious for their centuries-old exploits. The Stait is narrow and shallow with many small islets and river outlets. This makes for easy pirating. Even now Malaysia, Indonesia, and Singapore battle modern-day pirates.

Singapore now dominates the straits and trade. Malacca's port silted up and modern ships are too big to dock there. Today all shipping in Malacca is from offshore anchorages. Their harbor facilities handle rubber exports, rice, and sugar imports.

Shirtless and covered in sweat, two men sit in the shade of an Amla tree as we drive into Dutch Square. The buildings in the square are all red. Old Christ Church was originally Dutch Reform, but is now Anglican. It's made of red bricks from Holland. All the other buildings are bricks with Chinese plaster painted red. There are remnants here from all the peoples who once inhabited Malacca. The Stadhuys from the Dutch, Queen Victoria's Fountain from the British, Tan Being Sweet clock tower from a Chinese tycoon, and St. Paul's Church from the Portuguese.

It's hard to imagine this town of one and two-story red plaster was once the major trading port in Southeast Asia during the 16th century. Then the lucrative spice trade of cloves, nutmeg, and mace flourished. There were Arabian, Indian, and Persian trade bases in the city.

In the 14th century, Paramesvara, a Malay king who was kicked out of Singapore by the Javanese, made this sleepy fishing village into a prosperous sultanate. Enlisting the help of entrepreneurial Chinese fleeing the Manchu rule, Malacca became a major stopping place for traders to take on food supplies and fresh water from the surrounding springs.

In 1511, the viceroy of Portuguese India, Albuquerque, eyed the wealthy sultanate and conquered it. This was the first contact Malaysia would have with Europe. Whoever controlled the Straits controlled the trade. The Portuguese made enormous profits from it over almost a two-century rule. The Portuguese were followed by the Dutch and British.

We notice the people are a pleasing mix of races. The Portuguese tradition of not sending women to their colonies resulted in the Portuguese marrying local women. The Chinese intermarried with Malay women and became known as the Peranakan or "staits-born Chinese." They were encouraged to do so with tax breaks and incentives. In contrast, the British discouraged mixed marriages, touting racial purity.

Religious buildings are found everywhere in the city. You can visit Cheng Soon Teng temple, the Javanese Kampung Kling Mosque, the Anglican Christ Church, and the Catholic Church of St. Francis Xavier within a few blocks. The Javanese Kampung Kling Mosque boasts Chinese eaves, Portuguese tiles, Corinthian columns, and a Victorian chandelier; a showcase of Malacca's eclectic culture.

We climb St. Paul's hill above Dutch Square and the river, where we visit the ruins of the Portuguese fort, a Formosa Fortress. Near it are the ruins of St. Paul's Church built in 1566. Originally it was called the "Chapel of Annunciation" by the sailor Duarte Coelho. In 1521, he built it in gratitude to the Virgin Mary for saving his life at sea. In 1566, the Jesuits took over the church and rebuilt it. The church once held the remains of St. Francis Xavier until they were removed to Goa, another Portuguese colony in India.

St Francis Xavier was canonized as a saint because of his Catholic missionary work in Southeast Asia. He's reputed to have converted more people to Christianity than the apostle St. Paul. His body has stayed so well preserved that its pristine state is viewed either as great mummification or a miracle. Reportedly, when they moved the body en route to Malacca, a finger fell off the hand and blood flowed from the wound.

We climb down the hill and head to Jonker Street. Like most of the streets coming off Dutch Square, Jonker Street is narrow and crowded. There are restaurants, tourist shops, and antique shops. In 1974, most of the tourist shops are hawking pirate souvenirs.

We stop in a restaurant and order satay celup (seafood, meat and vegetable skewers with peanut sauce and Hainanese chicken rice).

"We don't have any silverware or chopsticks. How the heck are we supposed to eat this?" Roger asks.

"Look around. Everyone's eating with their fingers," Pete notes. "When in Malacca do as the Malaccans do. Dig in."

"No fingers until I wash my hands," Gisela says. Americans eat with one utensil, a fork. For our European friends it's doubly hard because they use two, a fork and knife. They hesitate longer than we do before digging in. The skewers prove to be easy, but eating rice is messy. At the end we hold up our sticky hands and laugh at the rice stuck all over the table and our faces.

We find a small guest house near Dutch Square and drop into bed. Malacca has not disappointed us.

After another walk around, we head back to Singapore, a three-hour drive. As we wait in line to drive across the Malaysian border into Singapore, we're greeted by a giant billboard saying in three languages, "No drugs." Next to it is a black silhouette of a soldier with a gun aimed at a foreigner's head. We get the message. Singapore will arrest you for spitting on the street, so drugs are obviously, more severely punished. We're happy we're with the Minders, our conservative Swiss friends, and not Zee and Leanne, who regularly brought drugs across borders.

On arriving back in Singapore, we're greeted by our friends the Hawkins, who treat all of us to dinner at their house. After dinner we say goodbye to the Minders. They will fly to the USA the next morning. When they arrive in Los Angeles they plan to stay with Pete's family. We load them up with presents and letters to give to the Fischers.

CHAPTER 15

Sleeping with My Husband in a Penang Whorehouse

We leave Singapore early on April 3rd. Our timing is unfortunate since Hindus, Buddhists, and Muslims all celebrate holidays in Malaysia this week.

Outside the train station is a sea of people which parts as we push inside the building, using our large packs as battering rams. Amid chickens, shouting, vendors, and throngs of travelers, Pete leaps onto the train, a pack on his back, and determination in his eyes. I follow laughing at the absurdity of the moment. Inside the train, Pete stretches out, covering two seats. I sit down, sweat dripping down my face.

"So much for third-class train tickets!" I yell at Pete. "Jumping on a moving train! It's crazy."

"Yeh, you're right. That was a close one."

With every stop, more and more people push into the car.

We need to go back to Kuala Lumpur (KL) to catch the train to Penang. By the time we reach Kuala Lumpur (KL), we're sharing our seats and floor space with four kids, a baby, and two chickens. The train from KL to Penang is as crowded as the train to KL.

Luckily, the transition from train to ferry is easy. We get off the train and walk onto the Butterworth Ferry, taking us to Georgetown on the Island of Penang. By this time, we've been traveling for fourteen hours. We're tired, hungry, and need a hot shower, but it's not to be.

"Take us to the closest cheap hotel," Pete tells the trishaw driver.

The driver wags his head and says, "Hotels full. No rooms. Many people. Holidays." Pete and I don't believe him and feel he just wants to extort us for a high price hotel.

"Take us to a hotel," Pete demands. We go to one hotel, then another, and another. Each one is grander than the last. At each one, the answer is the same: no rooms.

"See, no rooms," the driver says, smiling.

"You're right. Now, what do we do?" Pete asks

"I know one place. Room not cheap. I take you," the driver answers.

After a short drive, we enter a street where loud music throbs with a pulsing bass. Flashing neon lights shine above us. I wonder where we are, catching glimpses of shabbiness hidden underneath the glitter. We stop at a bar with a large red lantern hanging outside the door—our driver motions for Pete to follow him and for me to stay in the trishaw. Pete comes out several minutes later looking bemused. "OK, we've got a room for the night!" "Here? In this bar?" shock registering in my voice. He gives our driver a generous tip and leads me into the orange-lit bar.

A rotund Chinese man in traditional clothes waves us toward a darkened staircase. As we climb the stairs, we see several heavily made-up Malay girls wearing lingerie. They are draped over the second-floor railing looking and smelling like exotic gardenias. Although it's our first time in a whore house, it isn't hard for us to know where we are. The overly-made up girls, the red light, and the pimp are irrefutable proof. Pete leads me up the stairs wide-eyed and gaping as the women blow kisses at Pete.

Our room is windowless, eight feet by six feet, with a shiny, purple cloth covering the bed. The room smells faintly of jasmine and strongly of regret. Pete's grin widens as he whispers, "Guess we're not the first couple to sleep here!" A giant padlock hangs on the doorknob.

We lie down, trying to sleep. Below us, the music thumps, punctuated by laughter. Around us, we hear moans, grunts, pants, and cries. The bed is full of biting bugs, so we retreat to the floor.

When we wake up the next morning, we find the prostitute whose room we paid for, curled up asleep outside our door. The night ends up costing us four times what we would have paid for a hotel.

"So now we can add sleeping in a whore house to our list of experiences," I say,

"Do you think it was safe?"

"Who knows? There was a big lock on the door."

At a nearby coffee shop, a Peace Corps volunteer recommends the Eden Hotel outside the city in the Batu Ferringhi beach area. There we find one of the best rooms of our trip. Above a shop is a small apartment with a private bath and a bedroom surrounded by open windows that look out onto the crystal-clear waters of the Strait of Malacca. A white sandy beach rimmed with palm trees stretches off to the east and west. A cool sea breeze blows through the windows.

In the morning, after a much better sleep than the night before, we head for a nearby restaurant. There we meet an eighty-year-old British fellow traveling alone.

"My family gave me a plane ticket for an around the world trip. I've never traveled, so it's been quite an adventure," he tells us.

"That's some present! How's it going?" I ask.

"It's been hard at times, I'm not as young as I used to be, but I stay in hostels like you. Young folks help me a lot."

We're not sure whether to be amazed or appalled. We're in our twenties, and we feel the trip is taking a lot of stamina, strength, and brain power. We can't imagine doing this in our eighties much less alone.

"Well, I had a once-in-a-lifetime moment last night," Pete says. "I slept with my wife in a whorehouse!"

"Whoa, Pete! Don't give him the wrong impression. We needed a room because we couldn't find one in the city, so we paid for the whore's room. To clarify, I wasn't the whore wife," I add.

We laugh. Sleeping with his wife in a whorehouse will become a life-long story for Pete, "I've only been in a whorehouse one time in my life, and it was with my wife." I always need to add a few explanatory details. We relax on the beach and dreamed heavenly dreams in our room above the sea.

On our return to Georgetown, we stay at the Sydney Hotel hostel. Several travelers sit in the lobby drinking tea and telling travel stories. They tell us where to book the cheap ship to Lake Toba, Sumatra, Indonesia.

We buy deck passage on a freighter across the Malacca Straits to Medan in Sumatra, Indonesia, then we'll travel by bus overland to Lake Toba. In the pristine, air-conditioned shipping office where we buy tickets, they tell us the freighter will leave on April nineteenth or maybe the twenty first. We reluctantly, resign ourselves to being in

Penang for another five to seven days. We're disappointed at first, but find the city full of surprises.

The following day, we wake up to loud bass drums and chanting. When we look out our window, people pack the street. Women dressed in royal blue, pink, and turquoise saris with gold trim walk by carrying baskets full of small clay, milk jugs. Shirtless men wear saffron and yellow-colored dhotis (sarongs tied in front like loose trousers) with skewers through their cheeks. On their heads, they balance elaborate head ornaments. Women and men mingle in the middle of the street creating a colorful montage.

As the men pass, we see fifty large hooks pierced through the skin of each of their backs, being pulled on by men following them. The skin stretches outward as the men following pull on the white ropes connecting the hooks to the penitents' backs. After the ceremony, when the skewers and hooks are removed, there's no blood.

A man standing next to us dressed in a silk kurta suwaar (long-sleeved, hip-length shirt worn over loose trousers) with a wreath of marigolds around his neck says, "Hello friends, you look confused! What you are watching is the Tamil Hindu Thaipusan Festival Procession. It happens yearly when the star Pusam is at its highest point in the sky. It commemorates when the Hindu deity Parvati gave Murugan a divine spear so he could kill the evil demon, Soorapadmen, and his brothers. It honors our deity Lord Murugan, the destroyer of evil who represents virtue, youth, and power."

I point to the men passing with hooks hanging from their backs and chests. "This doesn't look like much of a fun festival to me. Those hooks and piercings look painful."

"During this festival, we seek blessings, offer thanks, and repay Lord Murugan for prayers he's answered. He answers our prayers if we pray to Lord Murugan to cure a family member or make our business prosper. We repay his blessings with holy sacrifice: dance, food offerings, and bodily self-mortification," the bystander tells us. The bodily mortification explains the reasons for the hook piercings.

We follow the procession to the Balathandayuthapani Temple. Colorful stalls line the way, selling flowers, food, and milk-pot offerings. Pairs of yoked water buffalos with painted horns and silver and gold filagree cloths over their foreheads led by young men dressed all in white dhotis start the procession.

As we get closer to the temple, people smash coconuts into the street to throw away bad luck from the past, and to purify themselves with this sacred fruit before entering the temple. Finally, we see a statue of Lord Murugan in a flower-covered, silver chariot. In the temple, we join the worshippers and place several small pots of milk and flowers at the altar's base.

Thousands push into the temple pressing tightly around us. Pete feels a hand reach into the empty pocket of his shorts.

He grabs the hand and yells, "Hey man, what are you doing? Robbing me?"

The small Tamil cowers and tries to get away.

All eyes turn toward Pete and me.

A chorus of voices overlapping each other, "No, you are mistaken, sir! No one would try to rob you in the temple!"

The would-be thief frees himself from Pete's hold and scurries into the street.

Pete grabs my hand, presses his other hand over the money pouch hanging from his neck, and motions me to do the same as we rush out of the temple. In India, we'll soon experience similar masses of people, which will always put us on guard.

The next day, we travel around the island on three different buses and discover an orchid farm and several batik factories. It's so hot that we retreat every day after lunch to sit below slow-moving fans and share travel information with other backpackers in the hostel's lobby.

I buy an ankle-length, cotton cheongsam (a body-fitting sheath) with a mandarin collar, a button knot on the right side, and no sleeves. It's comfortable and cool.

We wander the streets of Penang eating pineapple slices on sticks, small pancakes filled with coconut shavings and cane sugar, fried banana fritters on skewers, and Mie Goreng. Mie Goreng is yellow, fried noodles with different ingredients of shrimp, egg, tomatoes, or fish added to it. Each banana-leaf bowl includes some sambal, a rich chili sauce with coconut, lime, and peppers. Coconut water finishes off our meal.

On our walks, we notice bright red splats everywhere in the street. Most trishaw drivers have bulging cheeks. They are chewing red betel nut, a two-thousand-year-old Asian habit.

Penang is known as "Betel Nut Island." There are betel nut sellers on every corner. Pete watches a seller set up his small wooden stand. He's an Indian dressed in a faded white T-shirt and a dhoti. The dhoti wraps around his legs and waist just above his knees.

"You want?" the seller looks toward Pete.

"Ya, sila, how much?" Pete answers.

The man holds up ten fingers twice and adds, "ringgits."

Pete hands him two ten-ringgit coins. The seller spreads a betel vine leaf and dabs on some lime paste. Then he tops the paste with thin slices of the betel nut, rolls it into a bite-sized quid, and hands it to Pete. Pete shrugs and gives the seller a questioning look. The man shows him how to chew it in the side of his mouth. Pete pops it in his mouth and begins chewing. His eyes water, he makes a disgusted face and then spits a stream of red juice onto the street. His teeth, mouth, gums, tongue, and lips are bright red. He motions to the water bottle I'm holding. I hand it to him, and he chugs down the whole bottle.

The vendors on the corner watch him and burst out laughing.

"So, how was it?" I ask.

"It tasted like cinnamon and nutmeg, but it pulled out all the saliva in my mouth. I feel dizzy and high but energized," Pete tells me.

"I hear it acts like mouthwash, helps digestion, creates a sense of well-being, and is a stimulant like chocolate or a cigarette."

"I'll stick to chocolate and cigarettes," Pete replies.

However, the mystery of the red-splattered streets and chipmunk-cheeked trishaw drivers is solved.

We debate going to the Snake Temple three kilometers from Georgetown. My husband loves snakes and collected them in high school, but I don't like them. I grew up in a neighborhood of boys who would tease me endlessly with the serpents if they knew I was scared of them. Pete is excited about visiting the Snake Temple. I'm not.

The temple was built in 1805 to honor Chor Soo Kong, a revered Buddhist monk known for numerous miracles, his healing powers, and providing sanctuary for snakes. Supposedly, soon after, the temple was built. The snakes arrived. However, many poisonous vipers twine the snake branches on the temple columns; the temple brochure says that burning incense makes the snakes harmless. I don't believe it.

"Maybe they'll let us handle the snakes or get a picture with them," Pete says enthusiastically. I carry a fistful of burning incense as we climb into the temple. I slow my pace. Where are the warning signs?

As we're steps away from the altar writhing with snakes, I see signs saying, No Touch and No photographs. I read them out loud for Pete's benefit.

We end up having no close encounters with snakes. Later I learn, they remove the temple snakes' venom glands but leave the fangs. I'm relieved when we move away from the snake altar to the inner courtyard, where we see two magnificent dragon-shaped pure water wells and the famous, giant brass bells.

On our last night in Penang, we take a trishaw ride around the city. Our guide gives us a thumbnail sketch of its history. In 1756, British Captain Frank Light made a deal with the Kedaj Sultan and changed Georgetown from a pirate port into a thriving free port (a zone where there are no or reduced tariffs on imports). In 1876, Georgetown became a British Colony. In World War II, the British abandoned the city without firing a shot and evacuated all the Europeans. The Japanese massacred thousands of primarily Chinese residents and forced the Chinese women to become comfort women for them. It was the first city liberated by the Allies. In 1957, Penang gained independence from the British Empire and became part of Malaysia but fought for and retained its free port status.

We ride in the trishaw by some of the vestiges of Georgetown's past. To the left, we see old Fort Cornwallis's red brick ten-foot-high walls from 1786. On Esplanade Road, we pass the city hall built by the British in 1903. It's a white mass of arched windows, columns, domes, and a grand entrance. Next, we pass Goddess Mercy Temple, a Taoist temple built in 1728 and originally dedicated to the sea deity, Mazul. Finally, on the edge of the Tamil Muslim neighborhood, we marvel at the Kapitan Keling Mosque, built in the nineteenth century by Muslim traders.

Campbell Street's neon lights soften in the night. In 1974, Penang was a quaint melting pot of cultures. We ride up into the Penang Hills. A cool breeze comes down off the green hills, and the straits reflect the soft lights of a city yet to have a skyline of monster-high buildings. Our last night in Penang is memorable. We'll remember this city fondly. Early the next morning we catch the freighter to Sumatra.

CHAPTER 16

Lake Toba, Sumatra's Forever Paradise

It's dawn, Sunday, April 21st. Penang slowly disappears in a veil of mist as we head into the turquoise waters of the Malacca Strait. We're sitting on the deck of the freighter with other backpackers headed to Sumatra. The freighter is more barge than ship. The upper deck is long and flat with half of it piled with boxed merchandise. The steering cabin is in the stern and rises five feet above the deck. The passengers have the other half of the deck. A flimsy tarpaulin covering woven mats has been set up for us.

"Are you headed to Lake Toba?" I ask two lovely French girls.

Their dark-made-up eyes scan me critically, "Oui, we tired of Singapore so we took the train to Penang for a nice vacation at Lake Toba."

"What do you do in Singapore?" I ask.

"We're escorts for businessmen who come into the city. It's easy and pays a lot, but the Singapore authorities don't like it. They're always giving us trouble about our visas," the taller girl, Jacqueline, replies. Her answer solves the mystery of my three-day Singapore visa.

"How long have you been traveling?"

"We've been away from France for three years, or is it longer?" Jacqueline says as she looks at her friend, Celine.

"No, it's been at least four years," Celine answers.

"That's a long time You must have saved a lot of money for so long a trip," I naively reply.

"Of course, we are women. Money is no problem," Celine says as she takes a long drag on her cigarette and smiles. I connect the dots.

The weather changes to blazing hot then cold and wet. The tarp covers part of the deck, but the sideways bursts of wind and rain come in on us and we get little sleep.

At dawn, we wake to low tide along the mangrove coast of Sumatra. The shore is littered with garbage, smelling of mud flats, and swarming with sea birds. Beyond the swamp the city of Medan rises up dark and shrouded in gray smog. Although we're still several miles away, we can hear loud city sounds of traffic and industry.

"We obviously haven't arrived in paradise, yet," I say to Pete.

"Yeh, this is pretty nasty, but Medan is one of the largest cities in Indonesia. It's what you would expect——noisy, polluted, lots of traffic and crime. Let's just spend the night and get outta here. Lake Toba is going to be better."

We watch as the dock hand throws out the bow line and forgets to hold its end.

"This crew can't even dock this tub. I don't think we'll be getting off this boat any time soon," I say.

"I wonder if they'll unload the cargo before we get off," Pete wonders. The crew appears with sacks of rice on their backs and go down the gangplank to pile them on pallets on the pier.

"There's your answer, Pete," I say as I point to the workers.

Pete tries to go down the gangplank, but is stopped by the only guy not working and wearing a shirt.

"No go. Wait for immigration."

The slow-motion lassitude of this humid clime is even more evident when it takes the immigration agent two hours to arrive. Time is not a priority in Indonesia.

Once ashore we flag down a becak, a two wheeled motor rickshaw. After much bargaining, he takes us to a hostel near the port and bus station. We retreat under the fan in our room, victims of this chaotic, slow-motion city.

After a seven-hour bus ride through lush hills and rubber plantations we arrive at Parapat just as the sun dips below the horizon, painting the sky in purples and gold. There before us is Lake Toba. Lake Toba, Sumatra is the world's largest crater lake, thirty miles long and ten miles wide. It was made from a caldera carved out by a super volcano over seventy thousand years ago. The volcano is still active and sometimes shakes things up in northern Sumatra. Toba is the largest lake in South East Asia and one of the deepest in the world. In the middle of the lake is Samosir Island, larger than all of Singapore, with over two-hundred-square miles covering more than half the lake. The Batak people revere the lake as their mythical home.

The following day, we take a three-hour ferry ride to TukTuk, a traditional Batak village on the peninsula of Samosir Island. I sit on a barrel on the deck, talking to my fellow Indonesian passengers using the little Indonesian I learned in Jogjakarta. I have a pretty good fifteen-minute talk until my Indonesian runs out. When they start excitedly asking me questions, I can't answer, but smile. Speaking even a little of the language breaks cultural barriers.

As we leave the ferry, a young Batak man wearing a headband and an intricately woven shoulder cloth spreads his arms wide, smiles, and says, "Welcome to TukTuk! Do you want to stay at my Batak house? It's plenty big. You will like! My name is Apul Gultom. What are your names?" Every time we get off transportation we're greeted by a barrage of poorly spoken English and overly friendly vendors which we recognize as a hustle, but Apul seems genuine, and his smile is irresistible.

"I'm Pete, and this is my wife, Kris. How far away is your house?"

Apul gestures toward the thatched roofs, his voice full of pride: these are our bolon houses. Each one holds the spirits of our ancestors protecting us and guiding our lives. Toba Bataks have inhabited this island for centuries." Aku opens his arms wide and then points up a small hill. "We are the largest native group in Indonesia--much different from the Javanese and Balinese. We still believe in the spirits of ancestors, rocks, weather, and plants here and talk to them."

At the top of the hill, we see several rows of Batak clan houses made of wood and bamboo with thatched roofs. The homes have a rectangular shape with a saddle-shaped roof.

"We have five to six people living in our houses. As our guests, I only charge you five rupees each for the room," Apul tells us. The price is a bargain we can't refuse. We follow him to his house.

"Batak houses have lots of decoration and symbols. Each part of the house is a different world. The roof is home of Gods; below the roof, humans, and the bottom floor is death, where we keep our chickens, pigs, and goats." Apul leads us up a ramp to a front porch and through a small, carved, painted door. It is so low that at six feet, Pete has to fold himself in half to get under it.

"We believe you need to lower your head in respect to the house's owner," Apul explains. Again we lower our bodies, respectfully, going up the stairs to our room. It's a small cubicle with mats, and a square window cut out of the thatched wall propped open with a stick. We put down our packs and head to the local open-air market, careful not to hit our heads on the way out.

The market is teeming with people, but towering over the Indonesians are our friends from Australia, Deedi and Will. It's a chance meeting! We're in the same place at the same time! It's like a family reunion! They take us to Restaurant Mongoloid, where we feast on salad sandwiches and pisang susu (a sweet, strawberry drink.) The Mongoloid is well-known for its dinner buffet, and we return several times while we're in TukTuk. We catch up with our friends, telling travel tales and planning to meet in Kashmir in the North of India.

It is Will and Deide's last night in Samosir, so they offer us their rental place to stay. Their place is a two-story A-frame on the lake with a sleeping loft, private bathroom, kitchen, living room, and a porch. This space is much better at thirty rupees (two US dollars) a night than Apul's room.

That night we almost set Apul's house on fire with our mosquito coil. The room is hot, buggy, and small. It's hard to tell our enthusiastic friend that we're moving to a new place so we hire him as a guide to see Tomok and plan to meet him later that afternoon.

It's sunny, so we hike up the hills around TukTuk and pack a lunch. We pass fields and villages until we're standing above Lake Toba and TukTuk. The island and blue lake stretch out below us. In the distance, we see a waterfall cascading into the lake. We revel in the aloneness until we see eyes peeking out behind the rocks and bushes. We hear "hellos," and then we're surrounded by curious Bataks. We'll discover you're never alone in Southeast Asia, no matter how high you climb or how far into the countryside you go. Our last photo of the hike is a rainbow ending at Tuktuk, a perfect symbol for this idyllic place.

Later that afternoon, we swim across the bay to Tomok because it's only a short distance by water and a long walk by land. The day is hot, and the water is invitingly clear and cool. Holding our clothes above our heads, we sidestroke across the bay.

Apul is there waiting for us. He laughs as we shake off the water. "You swim across the water very funny Americans." He points to a grouping of hand-carved stone chairs in a circle. "In Tomok ancient Bataks gathered and talked out their differences. We still talk through problems no more fighting. Once we ate our enemies. No more. Talking better."

"Maybe the West can learn from Bataks," I say. "Less fighting and more talking."

Apul leads us to a food stall where we buy bowls of Mie Gomak, Batak spaghetti with homemade noodles full of andaliman pepper, a Northern Sumatran ingredient. This spice is lemony with a strong tongue-numbing kick.

The next day our bus trip back to Medan shatters the mellowness of our Lake Toba visit. It turns into the worst bus ride so far. The bus is old and alternately speeds up and slows to a crawl. As it slows, black smoke spews from its exhaust, until it stops completely. It will take two more buses and twelve hours to get to the city.

We wait on its roof, smoking Buddha grass and laughing while listening to Kosmos Blank and Robert Onion. We had shared a coffee with them in Lake Toba before the bus trip, but didn't know them well. They entertain us with stories of their travels.

Kosmos has messy hair down to his shoulders, a beaded vest covering his shirtless chest, bell-bottom jeans, and a bushy mustache.

Robert Onion looks like a Buddhist monk gone bad with a shaved head and comically big ears. He wears Tibetan prayer beads around his neck, and a Grateful Dead T-shirt tucked into an orange sarong.

"I met a white guy in Goa who was a dealer. I stayed at his place for a month. He lived like a maharaja in a beach house with twelve bedrooms and a swimming pool. Half-naked girls were wandering around all the time. He was always meeting with sketchy looking business partners. The dope was amazing, but I decided to leave when he showed me his stash of guns," Kosmos tells us.

"Man, I loved Goa, but people really got strung out there. I met a German guy begging on the beach. He had matted dreadlocks, his shirt was ripped and faded and he was barefoot with overgrown toenails. He came for a vacation, took some serious drugs, and three years later, he's still living on the beach," Robert adds. We take Goa, India, off our itinerary.

The second bus arrives but dies on the outskirts of Medan on a tree-lined boulevard. While we wait for our third bus, a marching band of girls dressed in red and blue uniforms goes by. They wave their white cowboy hats at us as they pass, looking more American than we do. We spend a sleepless night near the bus stop in a cheap, mosquito-infested room.

A large crowd surrounds the freighter back to Penang; we push our way to the gangplank showing our tickets as we go. Our return trip is full of stars and no rain.

Once we arrive in Penang the immigration officials give us cheap travelers only two-day visas and charge a questionable fee for their services. The government wants tourists, but only those that spend more money. Pete and I will leave for Bangkok the next day.

In 2005 when we were living and working in Washington, DC, I asked a traveler friend of ours, Jack Wheeler, where to go on our trip. "Jack, Pete and I want to go on a trip for our sixtieth birthday. You've been to almost every country in the world. Where would you suggest we go."

"Without a doubt, Lake Toba, Sumatra, Indonesia," he answered. "It's a paradise."

"Really? We went there in 1974. It was thirty-six years ago, and yes, it was a paradise." Remembering this conversation as I write this chapter; I can't believe that Lake Toba has remained a paradise for all these years!

CHAPTER 17

Land of the "King and I"

Twenty-four years ago, Maersk Shipping posted Steen Christian Fischer to Bangkok. It was a hardship post in 1950, so he left his new bride, Maryanne Bruun, and his year-old son, Pete, with his mother in Copenhagen. Now Pete is returning to a place he's heard about all his life; Bangkok, the fabled city of family legend. For years in his California home, he'd looked at exquisitely carved elephants on their buffet and the black and silver serving set with palm trees and sarong-clad dancers, souvenirs from his father's stay in Bangkok.

The train slows into the station. Competing smells of spice and rotting garbage greet us. Honking cars, police whistles, and vendors hawking their wares add to the chaos as we ride to the Malaysia Hotel. Our motorcycle rickshaw driver weaves skillfully around bikes, elephants, carts, limos, and trucks.

The hotel is a large six-story concrete structure with a swimming pool, TV room, and restaurant. In addition, it's near the US Embassy, the Joint US Military Advisory Group, Patpong red-light area, and a bus terminal. This air-conditioned oasis in the middle of Bangkok costs only sixty baht. A baht or tical is worth a nickel in 1974, so this hotel is a bargain.

Built in the 1960s, the Malaysia Hotel was an R&R destination for US enlisted men during the Vietnam War. When the war wound down in 1973, G.I.s were replaced by '70's hippies. Later the hotel would be the setting for the 1987 anti-war movie, "Good Morning, Vietnam."

Thailand is the land of "The King and I." We immediately head to see the Grand Palace, which housed the royal family. The Grand Palace has many buildings, halls, temples, courtyards, and gardens. It was the official residence of the kings of Siam (later Thailand) from 1732 to 1925.

It's near closing time. We're wearing shorts and short sleeves which is not allowed, but the ticket seller waves us in The outside gate looks like a military installation, with a long row of guns and other weaponry on a long wooden rack. A guard sits stone-faced, looking straight ahead with a machine gun cradled between his legs. The ticket taker punches our tickets with his super-long Thai thumbnail. We thought those long nails were just for looks, but obviously, they serve a purpose.

The palace was once a sealed city with intricate rituals, social status, and intrigues. The inner courtyards housed the kings' huge harems, guarded by combat-trained female sentries. "Way back then they had women warriors. Women fighters isn't such a novel idea after all," I note.

The main palace is the triple-winged Chakri Mahaparast. A mondop which looks like a Thai classical dancer's headdress, a layered, heavily ornamented point, tops each spire. The tallest spire holds the ashes of all the Chakri kings, and the two other spires hold the ashes of all the princes who never became king.

Land of the "King and I"

Our tour guide races us through the labyrinth of buildings. We'll have to return tomorrow to see the Reclining Buddha, Wat Pho, and the Emerald Buddha, the most revered Buddha in Thailand, in the temple at the Grand Palace. The intricacies and beauty of these buildings humbly remind us of the grandeur of Thailand's past.

Early the next morning, we set out, looking for a used bookstore a fellow traveler had said was nearby. Using the buses is easy because there's a bus stop outside the hotel. Book reading makes the long train and bus hours go by faster. Pete is a fast reader and likes detective novels and nonfiction. I'm a slow reader, and I like fiction books with settings of the places we plan to visit. Before we visited Singapore, I read King Rat, about Changi Prison. Our packs have limited space, so one book each is our carrying capacity. Pete has trouble hiding his irritation when I'm slowly finishing a book. "You're not going to have a test on this book. Read faster! I've finished my book. Are you memorizing every page? It's a good thing I like to reread books."

"Reading isn't a speed contest. Besides, I'm looking at the writing and the author's underlying themes."

"Oh, Rohver, seriously. The underlying themes? Just read and do it faster. Smart people can read fast."

"What? Are you calling me dumb because I read slowly? Reading fast isn't a sign of intelligence."

We move away from each other and then look around the store. I pick up Mitchener's <u>Caravans</u> and <u>Four Reigns</u>, a book about life in Bangkok's closed city. He's thumbing through a book by Jim Corbett about hunting tigers in India. At the register, I hesitate and then put <u>Caravans</u> on the counter. Pete puts the Corbett book down.

"We can buy both. OK?" I ask.

"That's too much money for both; besides, there's not enough room in our packs."

"Maybe if you hadn't bought that heavy world coin book in Singapore, we would have plenty of room and more money."

"That coin book is invaluable. I'm able to know what coins to buy. We'll sell those coins back in the States for a lot of money. They're an investment."

"You're not going to sell those coins? You're a collector, not an investor. How many coins have you sold in the past? I think NONE is the answer," I say, "Here, buy your book. I'm leaving!" I slam my books on the counter, stomp out, and slam the door when Pete grabs my arm.

"Where are you going? "

"Anywhere you aren't. I'm tired of you talking me into what you want and being such a cheapskate."

We both storm out the door with no books. Pete goes one way, and I go the opposite direction.

It's our first fight. I wander the streets of Bangkok crying and end up at a coffee shop where I spend a lot of money on an overpriced cup of coffee and cupcakes. Back in the room, I restlessly pace wondering when or if he'll return. Although it feels like an eternity, it's only a couple of hours since we parted. He opens the door, walks to where I'm standing and hands me a small wrapped box. He smiles sheepish and says, "Will you marry me, Miss Rohver? I never did get you an engagement ring. I know you don't like diamonds, but this is prettier. You aren't stupid, and I'm not a cheapskate. We just had our first fight. It was awful. Will you accept my peace offering?"

"Will you accept mine? I hold out a cupcake to him." I open the box, and a black star sapphire ring sparkles up at me.

"Oh, Pete, it's beautiful! I love the way the light moves across its surface. This ring isn't something a cheapskate would buy. I love you so much."

He folds his arms around me and hugs me tightly. "I love you Miss Rohver even if you read slowly."

I reach up and kiss his cheek, "I love you too, even if you read too fast."

The next day we go to the Maersk Shipping Lines office in Bangkok. There's a telex from Pete's dad who still works for Maersk, and now runs the Los Angeles office. We send one back to him telling him we're healthy and enjoying our trip. It seems incredible that in 1974 telexes were the primary communication for businesses.

Chatree Chomthavat, the head of the office, invites us to dinner at his home that night. He and his wife, Pringprao Boonyarataphan, who works for the Canadian Embassy, live in a modern house in the suburbs of Bangkok.

Thai people have impossibly long names that are hard to pronounce. They kindly shorten their names for us. We call them Chatree and Pring. Twenty years later, while teaching at the American Embassy School in New Delhi, I had a young Thai boy in my class whose name had fifty characters. I end up calling him Pong.

Chatree and Pring are successful, young, and charming. They both speak impeccable British English. Pring is stunningly beautiful in a silk sari. I'm glad I have my shimmering black sapphire on my finger, but Pete and I feel shabby next to this regal couple. Sitting on cushions at a low table, we look at a garden full of orchids and jasmine. We eat course after course of Thai delicacies. It's a magical evening that ends unexpectedly.

On our way back to the hotel, Chatree takes us to a high-end whore house. We walk into a large windowless room with a semi-circular glass wall behind a circular riser covered in vermilion velvet. Seated on the well-lighted risers are fifty lovely Asian women in scanty lace and silk robes. Around their necks are necklaces with a large number hanging from the gold chain. A desk with a staircase going upstairs is in the darker part of the room.

Chatree explains to Pete, "You go to the desk, give them the number of the woman you want, and then go to a room upstairs to meet them. Do you want any of these women?"

"No, I have my woman standing next to me," Pete answers. I stand awkwardly next to Pete shifting from one foot to the other. We're unsure what to say. This is a different culture and different morals.

Chatree smiles a Cheshire grin. "We want to treat you to all Bangkok has to offer."

"Thank you for the hospitality, but No. No thank you,"

"I'll have my driver take you back to your hotel then."

"Good night and thank you for a lovely evening," I say as we walk toward the exit. In the back of the limo we give each other incredulous looks. Pete's dad is an executive at Maersk shipping, and a faithful husband. Has Chatree read us wrong or is this typical of business in Bangkok?

Pete learns his old principal at Agoura High School, Al Acton, is now the headmaster of the International School in Bangkok so Pete gives him a call.

"Al Acton speaking."

Land of the "King and I"

"Hi, Mr. Acton, this is Pete Fischer calling. I was one of your students at Agoura High School in the first graduating class of 1967. I'm here in Bangkok and thought I'd give you a call."

"Yeh, yeh Fischer. You have a large family. I remember your brothers and sisters, Anne, Doug, and Barbara; right? How would you like to come out to the school at Sukumvit, Soi 15 today at noon? I'll treat you to lunch and show you the campus."

"Will it be OK to bring my wife?"

"Wow! Wife. It doesn't seem that long ago you were running down the school halls delivering audio visual equipment with Eric. Sure, both of you are welcome. It's not often I get Agoura alumni visiting."

The entrance to the school is impressive with gold metal lettering with the name of the school in English and Thai. A series of brick and beige concrete buildings make up the large complex.

Al greets us warmly at the entrance. "Class of '67! Those were three hectic years opening Agoura. We didn't have a track, gym, or auditorium, but there weren't a lot of students yet so I got to know all of them well. Wasn't your class the first graduating class.? About how many were in your class?"

"Only fifty of us." Pete answers.

We walk into the cafeteria which looks and smells just like a US cafeteria. We smile as we nostalgically breathe in the familiar.

"It was wild my first few years here. The Vietnam War had just escalated so our numbers soared to over three-thousand students. There were lots of CIA and officer's families here then. We ended up building a second campus across town. Since the war ended our student numbers have dropped."

"I bet it was challenging this close to the war," I say.

"The real challenge was all the fatherless households and easily accessible drugs. Our kids got in trouble. I did my best, but it was hard."

"I was lucky and pulled a high draft number so I was able to finish at UCSB. I'm glad I didn't have to go," Pete says. "The war messed up a lot of guys." We leave feeling the weight of our generation's war.

Our next stop will be Burma. It's a closed-off military dictatorship that discourages tourists by issuing seven-day visas and allowing entry only by flying in on a Burmese Airline flight. You need to first buy your plane ticket and then take it to the Burmese embassy so your visa can be processed.

While we wait for our Burmese visa, we decide to visit the Reclining Buddha, the Emerald Buddha, the snake farm, and the largest outdoor market, JJ Market. We also find time to send our moms' Mother's Day telegrams.

Wat Pho contains the largest reclining Buddha in the world. Unfortunately, the golden statue is almost too long for its shelter. The complex also houses a vast collection of Buddha images, a monastery, and classrooms where traditional Thai medicine and massage are taught and preserved. It is truly magnificent.

We cross the street back to the Royal Palace, where we make our way through the buildings to see the revered Emerald Buddha, which dates back to the 14th century. This building stands out in the royal complex with lovely mosaics of semi-precious stones and a mural of the Thai version of Ramayana.

High atop a golden wedding cake is the tiny Buddha, only sixty-six centimeters tall, dressed in gold monastic robes, which the king changes three times a year. As the vibrant chaos of Bangkok roars outside, the emerald Buddha's serene gaze stills my racing thoughts—a reminder of Thailand's enduring grace amidst its modern hustle.

The snake farm is a research center where they study the fifty-six venomous snakes in Thailand. Before the center opened in 1922, there were many tragic deaths from snake bites. They make anti-venoms for the King Cobra, the Cobra, the Malaya Banded Krait, the Banded Krait, the Russell's Pit Viper, the Green Pit Viper, and the Malayan Pit Viper. In 1974, it wasn't an educational center, so we wander around concrete snake pits where snake handlers walk through the enclosures, grab snakes' heads with their hooks, and milk their venom into plastic-covered jars to make the anti-venoms. I hate snakes, but Pete likes them. I take deep breaths as I watch the men walking among the snakes. It's scary and risky even watching from above. My tell-all face can never hide my feelings. He places himself between me and the pits shielding me from the edges.

J.J. market is a raucous weekend market with over 15,000 stalls. I photograph baskets of fried grasshoppers, beetles, durians, jackfruit, rambutans, mandarins, and custard apples—foods with tastes and textures both savory and repulsive. Pete looks for coins and paper money. I buy some $0.15 soap which they wrap in a now worthless one-hundred-yuan note from the Central Bank of China.

On our last night in Thailand, we go to a sex show. We have been given daily handbills for these shows and are curious. Once in Patpong, the red-light district, we're harassed by vendors selling their shows. "Biggest titties here!"

"Lots of beautiful naked girls!"

"Real sex act on stage. Must see!"

"No cover! Just buy drink!"

"Good dancers! Lots sexy! You see!"

We descend into a smoky basement and are seated at a table near a roped stage. Thumping bass booms as a naked girl appearing drugged sashays down the runway in five-inch heels. No stripper G-strings and nipple tassels needed in Bangkok. She is totally naked.

At the end of her performance, she rolls over on her back and kicks up her legs in a wide V. She flips over and then walks along the edge of the roped stage stooping low enough for the few rough men lining the roped area to hand her baht notes and touch her breasts. She slowly turns her back to us, and jiggles off the stage. We leave. Back at the hotel, we hug each other and go to sleep. The sex show is a turn-off.

After eight days in Bangkok, we have seen the best and worst of Thailand. Tomorrow we're off to mysterious Burma, the country we've heard about from so many fellow backpackers.

CHAPTER 18

Cigars, Jade, and Orchids

In 1974, Rangoon, the capital of Burma, is a decaying city in a stagnant country of oppressed people with little hope. Even the abundant green vegetation is garishly colored, growing atop buildings and escaping from cracked walls. All the structures are black with mold. Rusting machines stand idle in empty lots.

The only legal way into Burma in 1974 is to fly United Burmese Airways, possibly the worst airline in the world. The in-flight meal consists of a stale cheese sandwich and fruit juice, which we don't dare drink. Warner Montgomery, another 1970s traveler to Burma, explains it this way, "After World War II, Burma goes into isolation, only sticking its head out occasionally to check world affairs. In 1970 a military Junta cracks open the door to tourists offering a seven-day visa. To put it mildly, Burma was not ready for visitors in 1970."

Other backpackers at the Malaysia Hotel tell us to stay at the Rangoon YMCA but watch out for thieves. We take a taxi directly there from the airport. Remember there's no internet and no travel books for cheap travel in 1974. Our best source for travel information is other backpackers headed the opposite way.

In front of the YMCA is a thriving black market where you can exchange your dollars or merchandise at a higher rate than at the banks.

This is another tip we've been told by our fellow travelers. Four-and-a-half Burmese Kyats are worth about one US dollar. At the Bangkok duty-free store we buy two bottles of Johnny Walker Red Label and two cartons of Rothmans cigarettes, costing us thirteen dollars and fifty cents. We change four US dollars for appearance's sake, then sell the booze and smokes in front of the YMCA for $67 US.

Our finances complete, we head out to explore Rangoon. Friendly smiles and curiosity greet us as we walk down the streets. Squatting on every corner is a child or adult smoking thick cone-like cigars. It's one of the genuine bargains in Burma, costing less than a cent for reasonably good tobacco. We head to the train station to register for the morning train to Mandalay, a city directly north of Rangoon in the middle of Burma.

After we stroll through the night market with open stalls full of exotic fruits, gems, and handmade crafts. Lanterns give off a soft light and the ugliness of the city melts into the night.

"You want jade? I have," a merchant pulls at my arm and tilts his head toward his stall where all colors of jade lie on a wooden plank. "Lookey, lookey, Mickey Mouse!" he says as he spies my wrist watch. "You want to sell?

"I don't know," I say as I twist the red, plastic wrist band, but I unclasp the watch and hand it to him. The merchant's eyes sparkle as he holds the watch aloft.

"Mickey Mouse! My children will love this!" he says piling jade in my hands, each piece cool and luminous until I can't hold any more.

"Pete! Look what I just traded for the Mickey Mouse watch Kathy gave me. She was sure I could trade it for something special and I have." I open up my hands.

"What? Is that jade? For that stupid watch. Wow! I guess your sister was right, Mickey is worth more than I thought."

We sit down at a small stall and over a glass of too sweet tea count the pieces. There are thirty-eight pieces of jade. The stones are every color of green; milky green, dark green, gold-flecked green, and translucent green. What a deal!

We hail a bicycle rickshaw and travel back on streets lit only by propane lanterns and stoves. The city is almost lovely in the moist, velvet, darkness.

Before sunrise we're on our way to Mandalay. We go there instead of historic Pagan because our parents' collective memories of the 1939 movie, Road to Mandalay starring Bing Crosby, Bob Hope, and Dorothy Lamour have been implanted in us both. As we roll out of the station we sing "On the Road to Mandalay Where the Flying Fishes Play".

The train ride is slow and hot. Unfortunately, the only thing to drink is the world's worst orange pop. It's not even cold, but with the temperature at 100 degrees Fahrenheit, we drink gallons of it.

In Mandalay, we meet up with our friends, Zee and LeAnn, who we had met earlier on the cruise ship and in Bali.

"Well look who's here?" Zee calls out as we walk into the guesthouse.

"Nice to see some familiar faces," I say.

"How's your trip going?" Pete asks.

"Groovy. We flew from Bali to Bangkok and then tripped out in Kosa Mui and Phuket. Huts on the beach were only a dollar and the grass was amazing, What have you been up to?" LeAnn asks

"We traveled through Java, flew to Singapore, took a road trip along the east coast of Malaysia, then back to Singapore, Penang, Lake Toba in Sumatra, and then flew from Bangkok here," Pete tells her.

"You guys are really gassing it," Zee says. "We avoided Singapore. We heard it was a real tight ass police state."

"You're OK there if you stay between the lines, but I know how you guys like to stretch lines so it's probably just as well you didn't go," Pete continues, "We're getting up early tomorrow to look around before it gets too hot. Wanna come?"

"Sure," Zee replies.

We wake up early to climb Mandalay Hill using the western stairway close to our guesthouse. Two fierce lions guard the entry to the stairs. Around the base of the stairs is a cluster of stalls selling handmade opium weights, rice-paper parasols, gongs of varied tones, jade, sapphires, rubies, wooden horse puppets, wooden and bone carvings, belts, rugs, flowers, and incense. Everything is cheap.

After climbing the thousand steps we're rewarded by a panoramic view of Mandalay, the Sutaungpyei Pagoda, and several monasteries. Mandalay Hill has been an important pilgrimage site for Burmese Buddhists for two hundred years. We see one or two orange-robed monks walking around under paper parasols but no one else.

The stairs and pagoda are in disrepair. On our climb down, we notice hundreds of marble slabs inscribed with Buddhist scriptures at the foot of the stairs. Written Burmese is a magical series of curlicues and swirling lines that seem more like art than writing. Spoken, the language, is a mix of unfamiliar sounds and tinkling tones. No one can translate for us, so much of what we see at the pagoda remains a mystery.

Although it is already getting warm, we head to the Mandalay Palace near the base of Mandalay Hill. Surrounding the complex is a wide moat of dirty water choked with water weeds. We cross a poorly maintained bridge. Outside the walled city several young Burmese men in plaid sarongs and white dress shirts advertise guide services.

Our guide, Aman, tells us the history of the palace in his tingly-toned English "The palace was built from 1857 to 1859, the last royal palace of the previous Burmese monarchs, King Mindon and King Thibaw. The former royal palace of Amarapra was dismantled and moved by elephants to the foot of Mandalay Hill after the disastrous Second Burmese-Anglo War of 1852."

"Imagine how many elephants you would need to move a palace and how heavy their loads must have been!" I wonder out loud.

Aman continues, "In 1887, the British conquered the Burmese and plundered the palace, renaming it Fort Dufferin. In World War II, the Japanese stored their ordinance on the palace grounds, and, consequently, the Allies bombed the palace complex—lots damage."

Much of the wall is rubble. Only the watch tower and the royal mint are left.

We climb up the spiraling seventy-eight-foot watch tower. At its top is a seven-tier pyatthat (layered spired roofs). Mandalay spreads out below us. We wonder what this one-thousand-square-acre complex once looked like. How amazing it would be to travel back in time to see this rich culture in all its glory.

At noon it's unbearably hot, so we retreat to the awning-covered restaurant near our hotel where we drink strawberry susus (a smoothie made of condensed milk, milk, strawberries, sugar, and ice) all afternoon.

Sporting a graying ponytail, Forty-year-old Zee wearing a gauzy tunic shirt and frayed cargo pants settles his lanky frame into a low-lying lawn chair. LeAnn is in her twenties, with blonde hair going midway down her back. She wears a flowy skirt reaching her ankles and an embroidered top.

So many backpackers we meet traveling in 1974 are on a drug trip of discovery and enlightenment. Pete and I are not drug users. We're looking for adventure, a different kind of enlightenment.

"Look at this opium!" Zee says as he takes a large paper packet out and opens it. "All pure and so cheap. Too bad we only have a week. I could live here a lifetime."

He puts it away, and Leanne hands him a joint and adds, "We'll never be able to use all this. Do you want some?"

"No, but thanks," Pete replies.

Zee and LeAnn regularly cross borders with a stash because they worry about running out. This idea terrifies Pete and me. We never buy drugs much less cross borders with them.

Pete calls out to a tall, dark-haired traveler who has entered the restaurant, "Come join us!"

"Hello, I'm Andrew," he says, laying his pack by the chair beside him.

"Hi, I'm Pete. This is Kris, LeAnn, and Zee," Pete introduces us. Andrew is dressed like an 19th-century British explorer with khaki shorts, a collared shirt, and a pith helmet. Andrew hangs the pith helmet on the back of his chair and leans into the chair as he wipes his brow.

"Plenty hot enough," he says.

"It sure is. We just climbed Mandalay Hill, but I think we'll have to retreat to our guest house now until it cools down," I reply.

"I haven't walked around much yet, but I was planning on it. I'll probably do it later when the sun goes down as well. I'm a botanist from Kew Gardens in London on a month's vacation. Tomorrow I'm headed up to Maymyo to check out the botanical gardens' orchids. I've booked a jeep and driver to take me to Maymyo early tomorrow morning. Do any of you want to come with me?"

"What do you think, Kris, Leanne, Zee? Do you want to go?" Pete asks.

"Sure! Tell me more about Maymyo, Andrew," I say.

'Maymyo is an old hill station sixty-five kilometers West of here in the Himalayan foothills. I'm here to see orchids at the Maymyo Gardens. Alex Roger, a forest researcher, started the garden in 1915. He went all over Burma collecting specimens. Kew Gardens in London inspired the design of the garden. Lady Cuff, an amateur gardener, helped him with the design. I can't wait to see it," Andrew answers.

"It sounds interesting. I bet it's cooler in the foothills," I say.

"No thanks. Zee and I have found our piece of heaven here, so have fun, you guys. We'll see you when you get back," LeAnn says.

"OK, Andrew, it looks like Kris and I are going. See you in the morning." Pete says as we head back to our guest house.

"Cheerio,"

We leave at six in a fog that gets thicker as we climb. It's only sixty-five kilometers, but it seems much longer because the road is winding and in terrible condition. If you stay on this road past Maymyo, you get to China. Maymyo is well known for its role in WWII. Many American and British flyers flew over "the hump of Burma" to bring supplies to the Chinese fighting the Japanese. Along the way are brilliant white stupas (cone-like Buddhist shrines) --lone sentinels blessing the lush green countryside.

In 1974 the British Hill station still carries the name of its founder Colonel May, Maymyo. (Maymyo is now known as Pyin U Lwin.) We see the British influence everywhere. Horse-drawn carriages line the main street; the Purcell Clock Tower stands in the center of town near the Episcopal church, and the Maymyo Botanical Gardens.

The houses are neatly lined up with their names attached to the iron gates; Briarcliff, Heathstone, and Mayfield. Stone walls surround the overgrown gardens of lavender, hollyhocks, and violets. Made of red brick and stone, the houses have stubbornly survived their British owners. We could have been taking a stroll in Dorchester or Chelsea.

In the morning, Maymyo sounds more international. From the local mosque, a muezzin sings the call to prayer. The clock tower tolls, church bells ring, Hindu Sadhus chant in the temple, and gongs resonate from the pagoda.

We visit the National Mamyo Botanical Gardens (now called the National Kandawgyi Botanical Gardens) The orchids shimmer in the dappled sunlight, their petals delicate silk radiating hues of mauve and pale yellow. Although the garden is overgrown and neglected, Andrew finds rare orchids, spectacular flowers, and native trees to tell us about. We couldn't have a better guide.

We ride the jeep back to Mandalay from Maymyo. It's been an interesting detour, and we promise Andrew we'll look him up in London.

We book the train from Mandalay back to Rangoon. Since we have a few hours to wait, we go to our favorite restaurant for more strawberry susus.

Cigars, Jade, and Orchids

The sky has been progressively darkening all day. The locals tell us the monsoon will break at any time now. As we're sitting at the restaurant, it does. Sheets of rain pour down. We lift our feet up on the restaurant railing as water streams across the floor. The Burmese rush into the street joyfully stretching their hands skyward. At last, it's come, life giving rain! The scorching hot and dry are over.

Men armed with heavy sticks and stray dogs wait expectantly at the end of the drain pipes. A rush of rats pour out the pipes. Growling dogs and shouting men descend on the rats with a mixture of glee and viciousness. The dogs fling the rats high in the air, catch them in their mouths, and shake them vigorously. Men smack the rats like baseball players going for a homer.

The awning over us collapses. It's a sign. We shoulder our packs and wade into the street. Before we reach the train station, we are thigh-deep in muddy, rushing water. A four-block trip takes us thirty minutes through the rushing water. We finally get to the station and board the train for Rangoon.

Only an hour later the strawberry susus take their revenge. Local ice and water are in the Susu. We have ignored traveler wisdom about never drinking the local water, and now we suffer the consequences. Rumbling starts in our stomachs, and soon we're in the train car's bathroom with gushing diarrhea. Surrounding us are squatting Burmese who have snuck on the train and hidden in the bathroom. This is the breaking point for me. Pete bribes the conductor, and we move to the first-class carriage with a private bathroom.

In Rangoon we say goodbye to Zee, LeAnn, and Andrew. Although Andrew invites us to London, we don't make it. We never cross paths with Zee and LeAnn again.

The week we spend in Burma ends up costing us $17.50, thanks to the black market. These markets creatively circumvent government oppression and corruption, highlighting the Burmese peoples' resilience in the face of financial hardship.

Tomorrow morning, we fly out of Burma to Calcutta and then to Kathmandu. We hope we're well by then, but vow to be more careful about all forms of water from now on.

CHAPTER 19

At the Top of the World

In the early morning darkness, lanterns flicker as we head to the airport. Rumors say that if you overstay your weeklong visa in Burma, the military will put you in jail, so we don't want to miss our flight! We're also worried about the vintage coins and jade we carry. Will Burmese customs find our treasures, confiscate them and detain us? To our relief, immigration and customs wave us through.

We fly from Burma to Calcutta, India, and then to Kathmandu, Nepal. We change our plans to go to India when we find out there's a national train strike and no trains running. Since we'd planned to use trains to get around India, we decide to go to Nepal instead. Our quick trip there ends up lasting two months. As a traveler you're at the mercy of Mother Nature and governments. There are no certainties, so you learn to pivot fast.

As we descend into Kathmandu we see the snow-capped Himalayas sitting on a bed of clouds rising above us. Below us is the green plateau of the Kathmandu Valley. The airport sits in the middle of the valley, with the old city only six kilometers away. We go through immigration and customs quickly in the small airport. Brand-new Toyota cabs wait to take us into the city outside the concourse. They have been given to Nepal by Japan. Our cabbie speeds off through the maze-like streets of Kathmandu. This ancient, medieval city with people in traditional dress feels like another world to us.

There's a jolt as the cab stops in front of the Kathmandu Guest House. These first-generation drivers have no idea of how clutches and gears work. The cars screech and grind through the streets. We wonder how many transmissions will fall out before the drivers learn how to shift gears.

In 1974, the Kathmandu Guest House is the best hotel in town. There's a garden, restaurant, and five-dollar rooms. We splurge after penny-pinching through Burma. New places energize Pete, but I take longer to adjust. I head to the room to recover from a day of three different countries, two time zones, drastic climate changes, and leaps in elevation. He bounces off with the camera. His theory is that you need to take lots of pictures early on in your visit before the sights become too familiar.

The next day, we look for a cheaper place to stay. We go to Freak Street, a narrow lane south of Durban Square. The road was renamed by locals in the 1960s when "strange" young travelers from Europe and the US started hanging out at the hashish cafes on the street. Kathmandu is full of backpackers. It's one of the ultimate destinations on the Hippie Trail because government marijuana shops sell legal, cheap hashish and marijuana. Walking down the street and inhaling makes you high.

At the end of the '70s, Nepal outlawed the sale and use of drugs, imposed a dress code for foreigners, and deported many of the hippies to clean out Freak Street. Today, Old Freak Street still exists in Kathmandu, but it's much tamer, selling souvenirs and Western food without the drug vibe.

Restaurants along the street cater to hippies, selling burgers, French fries, marijuana-laced brownies, and American apple pie when the "munchies" hit. Aunt Jane's restaurant and the Snowman Cafe are everyone's favorites.

An enterprising merchant hands us a complimentary golf-ball-sized chunk of hashish and his flyer. The Eden Hashish Centre promises the best Nepalese hash and ganja (marijuana) in Nepal. The flyer's bold print says, "Let us take higher!" We end up giving away most of the hashish before we leave. We aren't going higher with drugs. Traveling is our high.

Our new lodging costs us thirteen rupees, about one US dollar. We drop off our packs at the hostel and head to general delivery at the post office. Unfortunately, there's only a Hawaiian postcard from my sister, Joanie. In 1974, without the internet, it's hard to keep in touch with loved ones. I feel lonely and isolated at times without contact. Everyone thinks we're in Calcutta now so all our mail is at the American Express office there.

All our traveling money is in American Express traveler's checks, so if they're stolen or lost, the money is refunded. American Express will also hold mail for travelers. We have the Calcutta office forward our mail to Kathmandu.

We've known druggies who boast about selling their American Express checks. Then, they claim they're lost or stolen and get reimbursed.

Playing loose and lawless in a foreign country is not a good idea. If you're broke and addicted, it's time to go home! The American Embassy sends many out-of-cash, drug-addicted, young Americans home with an expensive first-class plane ticket, then confiscates their passports until they can pay back the cost of the fare.

On Sunday, we rent bikes and go five kilometers outside the city to Swayambhu stupa one of the holiest Buddhist religious sites. (a stupa is a spherical mound-like structure containing religious relics used in Buddhist meditation and worship.) At dawn, we watch the Buddhist monks with shaven heads, wearing crimson robes, circle the stupa counter-clockwise. They rumble deep-throated chants, turn the prayer wheels and finger their amber prayer beads.

Tin-covered shops line the street to the stupa. Tibetans sit cross-legged on maroon and gold-colored rugs, selling Tibetan coins, heavy jewelry, khukuri (Nepali curved knives) machetes, sacred tanka paintings, and clothes embroidered with the stupa's all-seeing eyes.

Swayambhu sits on a hill. The largest stupa sits on a large white dome topped by a cube-shaped base with four pairs of eyes on each side. Dark eyebrows make the heavy-lidded eyes look wide open. The third red eye between the pairs of eyes is the eye of wisdom. The eyes mean God is omnipresent. A question mark nose is between the eyes, the Nepali sign for number one. It represents the unity of all things existing in the world and the one and only path to enlightenment through the teachings of Buddha.

Above the eyes are thirteen tiers making up the stupa. Topping it is an umbrella. Around the stupa are stones inscribed in Tibetan and small prayer wheels. Two strings of prayer flags reach from the top of the stupa down to its base. One receives blessings with each turn of the prayer wheel and flap of a prayer flag. We climb the stairs, walk around the stupa's base, turn the wheels, and hang up prayer flags. You can't have too many blessings.

Swayambhu dates from the 5th century, so there are many temples surrounding it built by former kings. One is a Hindu pagoda built in the 19th century by the goddess Harati, the eradicator of smallpox. Swayambhu is known as the "monkey temple," but we see only two live monkeys. They perch on lion statues guarding the staircases to the stupa.

We hop on our bikes, searching for the monastery but only find endless rice fields and thatched adobe huts. Riding the mud berms separating the areas, we crisscross the countryside surrounding the sacred site with the snow-capped Himalayas in the background.

Robert Thomas, our old roommate in Bundeena, went through Kathmandu on his way to Australia. He gave us one hundred dollars as a wedding present to buy tickets on the Royal Nepalese Airline's sightseeing flight to the Himalayas surrounding Mt. Everest.

As Everest's snowy peak pierces the horizon, I feel the vastness of the world, and our small fragile place within it. The flight is as majestic and grand as Robert promised it would be. The Himalayas are the highest mountains in the world. Everest's summit rises above all the surrounding peaks. Snow covers the pinnacle, with wisps of snow swirling around its dark rocks. Cameras click as our fellow travelers, and we fly along the twenty mountains making up this part of the Himalayas. The sky is crystal clear as the sun rises behind Everest. All these mountains are giants, but at over twenty-nine thousand feet Everest is, as the Tibetans call her, "the Goddess Mother of the world."

As we sit in dingy restaurants smelling of incense and hashish, eating special brownies, and drinking chai, fellow travelers tell tales of walking the Kali Ghandaki River Valley to the border of Tibet. They convince us the trek is well worth the hike.

"You've gotta go. You feel like you're walking through Middle Earth the landscapes are so Tolkien-like," Gerhart the Anthropologist tells us.

"It's so easy. Just get a trekking permit, a map, and get on a bus to Pokhara," Freida says.

"The Mustang Valley just beyond Jomoson is a window into Tibet, a closed country full of mystery. The Tibetan refugees who've escaped the Chinese have a fascinating karmic hum about them," Gerhart adds.

"The round trip is one-hundred-forty miles which isn't too bad. Just make sure you have good shoes," Freida says and continues, "You don't see the real Nepal in the city. The mountain villagers will amaze you."

Nepal's monsoon begins in mid-June, so we need to get going because it's already May thirty-first. With permits secured and designs for our jade jewelry finalized, we set off for Annapurna's lofty heights and the lure of the Kali Ghandaki Valley.

As a last bash before heading into the mountains, we go to the US Marine barrack's party. Heineken beer is three rupees and a water buffalo burger is two rupees. (a Nepalese rupee is worth $0.10 US.) "Hippie Trail" word-of-mouth messaging has made this a popular Friday night destination. Young soldiers and backpackers mingle in a strange mix of crew cuts, long hair, uniforms, and jeans.

In the morning we take the bus to Pokhara, our starting point for hiking the Himalayas. Our expectations and optimism are high. We're oblivious to the reality of what awaits us.

CHAPTER 20

Trekking the Kali Gandaki River Valley

Although it is only nine in the morning, sweat is dripping down my face as Pete and I follow the path around Takuche Lake toward the Kali Gandaki River Valley. The Kali Gandaki River makes a deep gash through the Himalayas, separating Dhaulagiri on the west and Annapurna on the east. The river begins high in the Mustang Valley in Tibet and winds its way south eventually joining the Ganges.

Already the Indian army boots that I've specially bought for the trek are rubbing against my feet. I notice discarded soles along the trail looking suspiciously like my boots. Definitely, not a good sign, but how did I know three months ago when we left Australia that I would be taking a hundred-forty-mile trek to the Tibetan border and need hiking boots?

We walk on a muddy trail bordered by the lake and farm fields outside Pokhara. Nepalese farmers pause and stare at us. Our western clothing: beige shorts, t-shirts with faded lettering, and baseball caps set us apart as does our height. We're both vastly tall compared to these compact, muscular men.

I have a large bright orange pack on my back and Peter has its sister in British racing green. They look deceptively heavy, but we have lightened them of all nonessentials storing most of our travel books, extra clothes, souvenirs, and cosmetics back at the hostel in Pokhara. Only our down sleeping bags and one other change of clothes fill the cavernous space.

Our cab has brought us most of the fifteen miles to the beginning of the trail from Pokhara, so we feel confident that we can make the ten miles to the first village in the gorge easily. Since I'm sweating on this flat stretch, I worry about what will happen when we start climbing.

My feet and the heat are minor discomforts. I'm happy and enthusiastic taking in the lush green of the valley and every butterfly, bird, and tree. Years later, as I'm reading a collection of women's journal entries written as they traveled the Oregon Trail, I note how their early entries similarly rejoice about their natural surroundings and then shorten to pithy dark comments as they travel longer and farther across the country. I will have a like experience on the trek.

After an hour, a strange pair overtakes us. It's a wiry Englishman in immaculate khakis with an ancient, British pith helmet waving a butterfly net over his head. Walking alongside him is his young Nepali assistant. The assistant wears a daura suruwal (a knee-length sleeved shirt tied at the side in five pleats with a closed collar), pants, and a stiff brimless hat.

The old Brit and his assistant carry fully packed wooden crates on their backs. Even with this heavy load, they dance down the trail as light and free as the butterflies they're chasing. They stop at the base of the mountain and the collector beckons us into a mud hut for lunch. We sit at a rough wooden table with benches.

"Good day, I'm George Sedgwick from Sussex, but am now living in Kathmandu."

"Hi, Pete and Kris, travelers from the US backpacking around the world. Why are you collecting butterflies?" Pete asks.

"I came here on a lark six years ago to add to my personal butterfly collection and then last year the King of Nepal commissioned me to make a butterfly collection of all the butterflies here in Nepal for the National Museum. I get a small stipend and they've extended my visa," George answers.

Our meals arrive in two fully-filled bowls; one of dahl (lentils) and the other baht (rice). This will be our staple food on the trip. The dahl resembles boiled grass colored either institutional green or off-putting mustard with a sticky okra texture. For me, diet food.

"Eat up! The next few miles are a strenuous climb. You'll need this for fuel," George urges.

The following five miles are brutal, and Pete and I slip further and further behind the flitting butterfly collectors. We finally stumble into the village and collapse. Along the trail, I have left most of my lunch. Fueling up obviously didn't work.

"Hello travelers. I am Tyag. I have a nice guest house just down the street where you can eat and sleep," a wiry, older man says.

"We're Pete and Kris from the United States. Where did you learn your English?" Pete asks.

"I was in a Gurkha regiment for thirty years working for the British military all over the world. Now I'm retired and I've returned to my village here in the hills," he answers. He leads us to a clean room in his house along the trail.

Throughout our trip we've heard of the Gurkhas known for their bravery and fierceness. Field Marshal Manekshow is famous for his quote, "If a man says he's not afraid of dying, he is either lying or a Gurkha." In London, there is a memorial statue dedicated to the Gurkhas. Every Gurkha has his kukari (a forward-curving knife) front pleated trousers, and a chin-strapped terai hat on his head.

Travel legends abound about the Gurkhas. In Singapore, we hear a tale about an American special forces soldier who bet his pals he could get past the Gurkhas guarding the British Embassy. He thought he'd succeeded until he felt the cold steel of the Kukuri on his neck.

When the so-called Indian army boots which are in fact pea green high-top sneakers, come off, my feet are too swollen and blistered to walk further. It's only the first ten miles of our trip, and we need to stop. I angrily heave the boots off a ledge. I will finish the rest of the 140-mile trek in rubber flip-flops. After a day in the village, we go on.

Each blistered step tests my resolve, but as the dawn slowly comes up over Annapurna, I am struck by its majesty. The mountain glistens and shimmers in first light. Its chiseled and stark peaks stand out against the cloudless blue sky—a once-in-a-lifetime sight.

As the day continues, clouds swirl around the mountain, and the sky darkens. It is right between the dry and monsoon seasons. We're racing the weather to finish the trek before the rains start and the trail becomes impassable. Our slow beginning doesn't bode well for us making it back before the monsoon breaks.

The trail goes up and down like a roller coaster following the river. We walk by the river for half a day, where the vegetation is lush and tropical; then, we climb into a stark, rocky desert-scape, miles above the river.

One day we come across a large, wild field of marijuana. I perch on a boulder in the middle of the field, and Pete snaps my picture. Let me take you higher!

Another day we meet an entourage of Sadhus (Indian holy men). This trail is an ancient trading route between India and Tibet, and a sacred pilgrimage trail for the Hindu faithful.

The Sadhus are covered in white ash and carry Hindu tridents with silver bells. Their hair flows around their shoulders in tangled masses, and dusty white dhotis barely cover their bodies. Solemnly they walk barefooted and chanting as they scour the river bed for shaligram fossils (a long extinct, spiral-coiled shells) which they view as one of the five nonliving forms of Lord Vishnu. They barely lift their heads to look at us. They're headed to the source of the Kali Gandaki River, which conjoins with the Ganges at the Tibetan border, a most holy place. We pause silently, respectful of these deeply faithful men.

Regularly on the trail porters pass us carrying refrigerators, sofas, and bags of cement. Just for show, we're glad our packs look like a heavy load.

As we climb up, I can't resist drinking from the clear stream beside the trail. Peter warns me, but I'm thirsty, and I can't believe anything harmful could be in this clean-looking water so high up on the mountain. Shortly afterward, we go through a village further up on the mountain. I watch people bathe, wash clothes, and relieve themselves in the stream. So much for clean water. Soon after I get a bad case of giardiasis, a form of diarrhea which will send me rushing off the trail my insides churning.

One of my worst memories is climbing downhill through a rainforest alive with leeches. When I squat off the path, the black slimy parasites attack me. Swearing loudly, I pull them off leaving pools of blood behind. At the end of that day, we find these nasty hitchhikers tucked away everywhere on our bodies.

A macabre scene greets us the following day as we walk along the dry riverbed. A procession of solemn porters and an American couple pass us, heading in the opposite direction to Pokhara. She is in the lead with a wooden staff, dressed in a flowing black kaftan with her blonde hair streaming out behind her. Above, the vultures circle. Behind her is her partner slumped in a basket on a porter's back. Brown, scraggly dreadlocks fall from his bowed head.

"Namaste. Are you OK?" I ask her.

"Namaste. No, we're not OK. We've been meditating in the Mustang Valley near Tibet, and my boyfriend is having painful stomach cramps, hasn't been able to keep anything down for three days, and is having trouble breathing. It might be due to the high altitude there. We tried to get a helicopter in Jomson, but couldn't afford it. In Pokhara, we'll take the plane to Kathmandu," she answers quietly.

"Good luck. Can we do anything to help?" I add.

"Thank you, no, we just have to keep going. I'm hoping we reach Pokhara early tonight," she says. This encounter makes us realize how vulnerable we are in this remote place. When we return to Kathmandu, we visit him in the hospital where he's recovering from altitude sickness.

The trek is not well traveled in 1974. Some days we don't meet any Westerners at all. The few we do meet are memorable. Yani, a twenty-year-old from Yugoslavia, who we met earlier in a Kathmandu coffee house is walking to Dhaulighiri alone; intrepid and independent as ever.

Trekking the Kali Gandaki River Valley

Christian, a gentle German with a flute, walks with us through several villages. Like the pied piper, he pulls out his flute, starts playing, and the locals gather to listen and join in with their instruments. What a universal language music is!

A strapping Japanese man speeds down the trail, trying to do the trek in half the time recommended. An hour before dawn, he's throwing his pack on his back and striding off. We wonder if he ever stopped to look at the snow-capped wonders around him.

Although we're only twenty-four, the up-and-down climbs are killing us. I fall further and further behind Pete and just when I want to give up, there's Pete waiting around the corner. I sit on a large rock, tears streaming down my face, "I can't make it, Pete. I can't go any further. Remember in Adelaide when you said you didn't think I was tough enough to travel with you? Well, I'm not."

"I'm tired too. You can make it. You're definitely tough enough, Miss Rohver. Keep going I'm sure it's not much farther to the village. Here take the last of the juice," Peter urges.

In the village the local Sherpas (high altitude mountain guides) have scrounged canned food from the base camp of a mountaineering expedition from Spain. We buy the can labeled soupa de carne. The twenty-five cent meat soup is a bargain and supplements our diet of dahl and baht giving us much needed energy for the next day's walk.

We pass through Hengza, Suilkhet, Yang Di, and Lumlie without stopping. At Biretanti, we stop and soak in the hot springs outside the village. It's the Nepali equivalent of a Scandinavian sauna. We hop from the heat of the hot springs into the cold river.

At one point we almost get lost where several converging trails cross. Our trekking map is no help so we use common sense; following the river and well-worn track. We end up in Ghoripani, the highest village on the trek. From this vantage point, we can look back at Pokhara and the lake as the late afternoon mist wraps around us.

We continue to head north to Jomson. On part of the trail, we traverse a curving three-sided tunnel carved out of the cliff. One thousand feet below us, the river rushes by. One false move and we'll be in the river. Stooped slightly so our backpacks won't hit the cave ceiling we follow the cave for half a mile. We are Bilbo Baggins traveling to Middle Earth.

One day we come across a British agricultural station trying to diversify local crops with a hillside full of ripe peaches. I remember sweet peach juice dripping down my face as we enjoy our first fresh fruit in days.

On another day, villagers urge us to join a wedding. As drums beat and horns blare, the villagers welcome us into a kaleidoscope of red dye, blessings, and joyous chaos. At the groom's house, we sit on carpets in a circle. The priest and newly-weds work their way around the circle while each wedding guest puts sticky, orange rice on their foreheads, boughs of hemlock at their feet, and offers them blessings. Although we don't understand any of what is being said we add our blessings. The night ends with Arak wine and a rich meat dish. Unselfishly, the village has shared their joy with strangers. Their openness and hospitality leave a lasting impression.

One clear morning, the swirling clouds part so we can see Dhaulagiri, the 7th highest mountain in the world, rising above us in all its glory. Its name in Sanskrit means dazzling, white, and beautiful. It is.

A stone monument under its shadow soberly reminds us of those who've died trying to summit it. K2 is its sister peak beside it. Annapurna rises across the Kali Gandaki River Gorge from it. Humbly, we walk through the giants.

Rain begins to fall in the afternoons and the river grows in strength. Two of our fellow trekkers underestimate its power and try to cross. One of them loses everything so we give him two hundred rupees to get back to Kathmandu. The remoteness and magnitude of the river and mountains remind us how vulnerable we are.

As we climb the valley getting closer to the Tibetan border, the air becomes clearer, and the landscape changes to yellow wheat fields and pine trees. Medieval-looking adobe villages cling to the valley walls. Tibetan women climb upward to the towns with large sheaves of wheat on their backs. In Marfa, we visit a recently built Tibetan refugee camp. In a new tin-roofed building, men and women sit in a circle, singing and weaving carpets with their butter lamps lit.

In 1949 the Communist Chinese invaded Tibet, sending the Dalai Lama fleeing and persecuting the Tibetans. They cross into Nepal in record numbers with only a few meager belongings. In Marfa, we see men building a foundation for an army encampment. Later, we pass many Nepalese Army troops headed to the border, going the same way we are. This deployment of Nepalese soldiers will be the early beginning of Nepal's resistance to Chinese encroachment on its border.

In Marpha, we stay in an adobe dormitory with a large flat roof. I have a high fever and am exhausted.

"Kris has a high fever. Does anyone have some aspirin?" Pete asks among the travelers in the dormitory.

"Here's some aspirin, but the best thing for her to do is get naked in her sleeping bag and try to get the fever to break."

"Did you hear Kris?" I take off and throw my pants, underwear, sweater, and shirt out of my sleeping bag hoping the fever will break.

Pete's whispering wakes me up. "Kris, I keep thinking of you naked in your sleeping bag. It's driving me nuts. Let's go up to the top floor. I'll climb in your bag, and we'll bust that fever."

"I'm OK now," I answer.

"Come on. When will you get a chance to see this many stars again?" Pete pleads. We climb the wooden ladder to the hostel's roof and make love under a brightly lit tapestry of shooting stars and planets.

The next day dawns rainy and gray. I wake up to Pete standing over me with a cup of tea on the roof of the hostel.

"Let's get going before it starts to rain. We're so close to Jomson and the Tibetan border. I can't wait to see Tibet," he says as he nudges me with his foot.

"I can't, and I won't," I answer sharply. I shrug out of my sleeping bag and go down to the bathroom. When I return, Pete is seated on a bed in the dormitory with his coat and backpack on.

"I'm still sick, I'm also worried about getting sicker and the monsoon trapping us in the valley. You saw all those troops headed to the border. There's going to be trouble, and I don't want to be near it. Go if you want, but I'm going back to Pokhara right now."

I stuff my sleeping bag into my backpack, sling the pack over my shoulder, and stomp out of the hostel heading back down the valley.

Before I'm even out of Marfa, there's a tap on my shoulder. It's Pete holding my forgotten money bag and wearing a sheepish smile.

"I didn't realize how sick you are. Sorry. I thought you might need this. You left in a pretty big huff."

Trekking the Kali Gandaki River Valley

"Yeh, I was angry and not feeling well, but it's crazy for me to leave all my money and passport behind. The rain's coming down pretty hard. If the river rises much further, we might not pass through some of the riverbed stretches. Look how muddy and fast it's moving now," I say.

"Hey, the Nepalese are in and out of this valley even in the monsoons, so we'll make it," Pete reassures me.

"I don't think we're even close to being as tough as these people," I respond.

"You're right. We may have to pay the porters to carry us out," Pete smiles, and we laugh.

"You also left something else important behind, me." Pete wraps his arms around me, and we walk on in silence.

In the next town, we meet up with Kerry and Barry, our Canadian friends who started the trek the same time as we did. They dose me with some tetracycline, and my diarrhea vanishes. We're amazed that we haven't seen each other earlier on the trek. Since the monsoon rains are starting earlier and earlier in the afternoon we leave before dawn. Trudging through heavy rain with plastic over you and your pack is miserable.

It's the last day of the trek. Pokhara is a day's walk away. We fantasize about hot showers and pastries. The mountains are no longer visible, with monsoon clouds covering them. We've gotten out just in time. The trail is muddy, and we barely look up from our feet as we walk.

A few miles from Pokhara, a vision comes out of the mist. It's a blonde schoolgirl complete with backpack, uniform, knee socks, and oxfords. We could have been on a street in London instead of coming down rock steps out of the mountains of Nepal. Is she an illusion? Where is she coming from? Where is she going? She never stops and floats by us, remaining a mystery forever.

The trek promised adventure, but we never realized how hard it would be. Having made it, we embrace the confidence it gives us.

CHAPTER 21

Lepers, a Living Goddess, and Rosie the Brit

We have been away from Kathmandu for twenty-four days on the trek. Before we leave Nepal for India, we need to get some shots and a visa. The jewelry maker has promised that the jade jewelry will be ready by the end of the week as well. Our second stay in Kathmandu adds new names and stories to my diary.

While getting shots for India, we meet Rosie, a British nurse living in Kathmandu who becomes a good friend. She invites us to a leper colony where she volunteers. We gather the donations of toys, food, and clothes the diplomatic missions collected to take with us.

When caught early, leprosy can be cured by modern medicine. However, once it progresses, the prognosis worsens, resulting in nerve damage that cripples the hands and feet and causes paralysis, blindness, and nose disfigurement.

In 1970 the medical community tries to change superstitious and religious thinking that leprosy is punishment for bad behavior in a past life. In the small villages of Nepal, people hide early symptoms of leprosy so they won't be banned from the village and sent to a leper colony. By the time their leprosy shows, it's too late for a complete cure.

We take a bus to the leper colony which is about a half hour outside of Kathmandu. As we arrive, the patients rush out to greet us. We give hugs, play games, talk to the lepers, and give them presents. Our acceptance of them brings smiles. Their smiles break through centuries of prejudice, teaching us the simple power of human connection.

It's heartbreaking to see young children with their parents--victims of ignorance and superstition. Modern medicine will prevent them from getting the disease, but they'll never lead a normal life because of their isolation due to their parents never having their leprosy treated earlier. It amazes me that in the 20th century, although we have cures for many diseases, people are still dying because of lack of information and harmful beliefs.

A few days later, Rosie invites us to the British Embassy Compound for lunch. It's one of the best missions in Kathmandu. We eat off Wedgwood plates and puzzle over the six pieces of silverware. What cutlery do we use to eat Yorkshire pudding, roast beef, and custard?

To return the favor, we invite Rosie with us for the July 4th celebration at Kathmandu's American Embassy. Once past the gates, we're in middle America. Hotdogs, hamburgers, potato chips, pickles, ketchup, mustard, relish, chocolate chip cookies, cokes, and 7-Up are on a table covered with a red and white checkered tablecloth. It's been years since we've had American food.

People swim in the pool and play baseball, basketball, and volleyball. The American flag is everywhere. Surrounding the grill are men wearing polyester, double-knit, plaid pants, and orange and chartreuse golf shirts. The women wearing bell-bottom hip huggers with teased hair hold towels for tow-headed kids jumping in and out of the pool. We have been momentarily transported home, making us both happy and homesick. We stay until the last fireworks and the playing of our national anthem. Hearing the anthem makes us proud and grateful for our country.

As we leave, we're struck by the harsh contrast between the overabundance of the embassy compared to the scarcity and poverty of Nepal. Both embassies recreate miniature islands showcasing their affluence, and the United States and British presence. Beyond their walls, Nepalese sit by charcoal fires, thankful to have food and a set of clothes for each of their children.

The next day I meet up with Yani, the young Yugoslavian girl traveling alone who we've met several times. She and I are going to check out the Nepali National Museum.

At the museum entrance, two military men with guns stop and ask us to hand over our purses so they can check them. The guns are old and stand a foot above the small Nepalis.

"No!" Yani pulls her bag away from the guard's reach. "Why are you searching through our things? You have no right to do this!"

The guard steps back, surprised.

"Yani, they just want to make sure we're not bringing in anything to harm what's in the museum. In the United States, all museums ask to look inside your bag."

"Kris, what if they find things they don't like in our bags or steal something?" Yani clutches her bag tighter.

"We're tourists. They won't," I answer.

"You Americans are so trusting. But, in my country, we must protect ourselves from the government knowing anything."

"Let's watch them closely as they look through our bags. OK? I want to see the museum."

Yani reluctantly agrees. This incident makes me realize how much I take for granted the freedoms we enjoy as US citizens.

We enter the museum to see dusty glass case after dusty glass case of weapons, swords, shields, khukuris (Nepali curved knives), and armaments. At one time, the museum was the Arsenal Museum because it only housed weapons from Nepal's wars. The leather canon from the 1792 war with Tibet and Napoleon III's sword are the only exhibits that stand out. We don't stay long because there are only a few signs in English, and the weapons don't interest us.

That night I go to a play with Rosie, our English nurse friend, at the British Compound. I take the bike I've rented for the week. By the time the play ends, it's dark, around nine-thirty. As I pedal down the empty streets, I see several Nepalese police blocking the road ahead. I stop and hop off the bicycle seat holding the bike upright my hands on the handlebars.

One of the four policemen grabs the handlebars in front of my bike. I use the scant Nepalese I know to ask them why they stopped me.

"I'm going back to my hostel. Ke ho (why)?" I ask.

Their faces aren't smiling. I realize I'm in an empty street with four men alone and don't know the language or why they've stopped me. Terrified, my hands grip the bike handles tighter, and my heart pounds.

A second policeman steps forward and yells something I don't understand except for the word "woman."

"Sorry, I don't know what you're saying," I keep my eyes lowered and my voice meek. The four men tug my bike off the road and into a shady alley.

Will I be taken to a Nepali jail? Can't they see I'm a foreigner? I search for any word from any language that might help, but my voice is dry and scratchy. What will Pete do if I don't return to the hostel tonight? Can the Nepali police be trusted? Should I offer them money?

Tears stream down my face. Then I shout, "USA! USA! USA!" The loudness of my voice startles them, and they step back from the bike. There's a loud dispute, and then one of the policemen comes up, rattles my handlebars, and sternly says something I don't understand. He lifts his hands off my bike and waves me forward, saying more words I don't understand, but I do understand the motion. I get on my bike and pedal away quickly. I will never know what was said, but I knew it wasn't good from the body language and tone. I was lucky. That night incident makes me glad I'm not traveling alone.

When I arrive back at the hostel, Pete is waiting for me. He sees my tear-streaked face and wraps his arms around me. "You were supposed to be home at ten. I've been worried. Are you OK?"

"Soldiers or policemen stopped me. I don't know why. It was scary. None of them spoke English, so I didn't know what was happening. I didn't have my passport. I thought the police would take me to jail so I yelled 'USA.'"

That night I realize how different this trip might have been had I been traveling alone as a woman. I think and worry about Yani, and Haley although I know they're strong and confident women.

The manager of the hostel explains to us that Kathmandu has had a nine thirty curfew ever since the Chinese communists started coming across the Tibetan border two weeks ago and that's probably why they stopped me.

Our hostel is close to Durbar Square, and the next day we pass Kumari Ghar where the "living goddess" lives. Nepalis are carrying the goddess around the square in a procession. We decide to follow.

The "living goddess" is a deep-rooted belief of the Nepalis, both Hindu, and Buddhist, that a prepubescent girl is the host body of the divinity Goddess Durga.

The goddess reclines on a golden palanquin. She's dressed in red and gold with her face heavily made up. Her black hair is in a topknot, and on her forehead is painted a red triangle with the "fire eye," symbolizing her extraordinary powers of perception. She is barefoot. Her feet will never touch the ground until the deity vacates her body at puberty.

We've heard a lot about this goddess from travelers and Nepalis. The requirements to become a Kumari are similar to being a beauty queen in the West. She needs thirty-two perfections, such as a body like a Banyan tree, no blemishes, a neck like a conch shell, eyelashes like a deer, a voice as clear as a duck, and small well-recessed sexual organs. She must also be serene and fearless and belong to the Shakya caste of silver and goldsmiths.

The priests, the royal astrologer, and the king pick the divinity. After they perform several secret Tantric rituals to cleanse her body and spirit of past experiences, she starts her reign. The Kumari's divinity status ends when her menstrual period starts. It must be hard for her to transition from a goddess whose slightest wish is granted to a mortal again.

There are several Kumaris throughout Nepal, but the best known is the Royal Kumari of Kathmandu. By seeing her, we have earned blessings. Although touching her feet gives you many more blessings, we don't.

Nepal has been a magical place. After spending two months here, it almost feels like home. Leaving and going to India both scares and intrigues us.

CHAPTER 22

Washing Away Ten Lifetimes of Sins

We leave Nepal by bus to Raxaul on the Indian/Nepal border. As we curl our way down the foothills of the Himalayas, we wonder what India will be like. Every traveler has had a story about India; some fantastic; some horrific. Below us, the flat, brown Indian subcontinent stretches from the Himalayas to the Indian Ocean.

This large expanse of level land has been India's curse and blessing. Conquerors have easily rolled across the country claiming it. Alexander the Great, Genghis Khan, Nadar Shah of Persia, the Ottoman Turks, the Portuguese, the French, and the British. India's diverse states reflect all these conquerors, each leaving a distinct cultural mark. This blend is evident in every corner of its plains from language to cuisine.

The bus stops at the Nepal/Indian border. We take a pony cart through the four custom checks. A Frenchman with us gets caught with marijuana, and we pay his five-rupee bribe.

Raxaul is a typical border town, dirty, hot, and chaotic. We take refuge at the Tourist Lodge and Restaurant which advertises itself as the "best arrangement for tourist" It has mosquito netting, a fan and is opposite the bus station. We have a too-spicy curry dinner which we can barely eat, but there's no other options on the menu.

"Maybe all these spices will kill any harmful bacteria," I rationalize as I eat tentatively.

"Yeh, says the woman with the cast iron stomach. Leave the chapatis (an unleavened flatbread) for me," Pete answers. "Let's not check out the kitchen."

The next day we travel nine hours on a bus and train to reach Benares. In Benares we plan to stay with our Canadian friends, Deide and Will, at the Moustache Varanasi Hostel.

As we exit the train station, masses of motor rickshaws mill around us. Each driver yells loudly for our business.

"I know best hotel. Get you there fast!"

"Cheap ride. My rickshaw best!"

Pete pulls out his map of Benares with the Mustache Varanasi Hostel marked on it and says, "Mustache Varanasi Hostel. Close to the train station. Best Price."

Since we know the name and location of our hostel, negotiating a price for the ride is easy. In a cloud of diesel smoke, we head off through the city.

Smoke rising from the burning pyres and fog obscure the city in an otherworldly haze. The Ganges River flows fast and muddy lapping at the steps down to the water. Hindus believe they will be reincarnated in the next life, striving to become a higher form or leave the endless cycle of death and rebirth altogether.

People in saris and dhotis bath in the river washing away their sins. Chanting and bells clanging carry across the water. We look on silently feeling the holy and sacred nature of this place.

Deide and Will greet us at the hostel. Having met in Australia, we've crossed paths with our friends several times during our trip. Deide excitedly shares what she's learned about the city. "Benares claims to be the oldest city in the world. It dates back to before Nebuchadnezzar captured Jerusalem and sent the inhabitants of Judaea into captivity. Before Greece and Rome flourished, Benares rose in greatness and prosperity. Imagine how long ago that was!"

"Ancient cities defy the ages outlasting men and their ambitions. Here we are yet another generation passing through. How insignificant our lives and times seem next to such a vast history," I say.

The next morning, we head out to explore. We can see and feel its vibrant underpinnings. The sidewalk is alive with people and activity. A barber sits on a stool shaving a young man. An elephant walks right near the edge of the street as the man riding him trims overhanging branches. A naked Jain, perched in a tree, prays. Cows roam everywhere, leaving their droppings indiscriminately. A sacred linga (Hindu fertility symbol usually made of stone and looking like a penis) pasted with orange rice sits on the corner.

The noise is deafening. Vendors hawk their wares, "jalebi, jalebi, jalebi." (overly-sweet pastries). Motorcycle rickshaws backfire. Loud Hindu music blares out of the open shops. The smell of jasmine incense mixes with that of camel dung and barbecuing satay. Our senses scream, "Stop! Too much! Too loud! Too strange!"

We look for refuge and a silk merchant beckons us into his courtyard and closes the door—Ahhh, quiet, peace. We sit cross-legged on carpets, sip chai, and buy several pieces of silk partly as appreciation for him rescuing us from the chaotic street.

We continue to walk the city following the river and its ghats (stone embankments) down to the river. Funeral pyres spew ashes into the air.

"Hold your breath. Be careful not to inhale Sahib Ram or Memsahib Pavan," Will jokes.

"I think these folks are hoping for a higher form of life than a twenty-four-year-old Canadian," Pete answers.

"Shhhh, you two. Show some respect," I say. "Pretend this a Christian funeral of someone you loved."

"This funeral is pretty noisy. I can't hear anything over all this racket," Pete replies.

Priests are chanting and singing loudly and the pyres crackle as Doms (outcasts who attend the fires) poke at the wooden pyres. The smell of sandalwood is heavy in the air.

"Hindus believe that death isn't the end, but simply a transition to something better," Deide adds. "Maybe it's more of a celebration and not mourning."

"If that's your father or grandmother, it would be sad and sobering to watch them slowly being consumed by fire. But, why do I only see men in white, orange-clad priests, and Doms around the pyre. Where are the mourners and no women at all?" I wonder.

Deide tries to answer my question, "Maybe cremation is more of ritual or celebration—not a sad ending to mourn."

Rows of beggars line the steps to the river. "Memsahib, Please." A beggar reaches up his arms to me. He knows the cremations remind us of the fragility of life and the inevitability of death. Charity in this life will surely be rewarded in the next. I place five rupees in his brass jar.

We walk along the river, seeing several men in white laying a body feet-first down on a pyre of several pieces of sandalwood and other wood. Ghee (clarified butter) covers the top of the pyre. A gray-haired man circles the body three times counter-clock-wise leaving the body to his left and sprinkling holy water as he goes. Stepping back, he takes a stick of flaming sandalwood and lights the pyre. The men and priests chant and pray as the fire consumes the body.

"You're right, Kris. This isn't the time to be telling jokes," Deide says. We watch in silence. Our thoughts turn to our own mortality and beliefs about death.

Dory-like boats line the river and we negotiate an English-speaking guide to take us down the Ganges. Will tells us, "There's a hostel tale that the Peace Corps daily tests the Ganges River water and finds it pure and unpolluted even though people bathe, urinate, and throw cremation ashes in it."

"That's pretty amazing, if it's true," I say.

Rashid, our guide, says, "The Ganges is most sacred. If you bath in the river, you can wash away ten lifetimes of sins."

I look at the fast-moving, muddy water and reply, "Although losing ten lifetimes of sins, is tempting, I'm not going in it."

"Neither am I," Deide adds.

"I don't have any sins to wash away," Pete says smugly.

"You're lying, buddy. We know you," Will says, "We'd better throw him in to save his soul."

From across the river, we hear loud shouting.

"You imbecile! Do as I say," a fat passenger dressed in an immaculate white kurta and dhoti says. He raises an oar over his head to hit the shabbily clad oarsman cowering in the bottom of the boat. The second passenger also in immaculate white grabs the oar from behind keeping it from hitting the poor oarsman. The negative actions of the fat passenger have been washed away by the sacred river.

We skim through Benares in two days, missing Sarnath and the engraved stupa where Lord Buddha taught his first sermon, beginning Buddhism in 528 BCE. Sometimes you don't even know you miss the best sights until you get home.

We'll travel to Agra the next morning. Pete and Will proudly brag about finding the cheapest tickets for the train to Agra, 3rd class unreserved.

"Pete, don't you remember the last 3rd class train ticket we bought in Malaysia? Third-class unreserved tickets in India can't be any better. Boarding this train is going to go badly!" I warn.

"Kris, it's fourteen hours on the train for only three dollars!" Pete argues.

"We talked the guy down from ten dollars," Will adds.

They're so proud of their savvy negotiating skills I let it go. Will has a trust fund and plenty of money, and although we're on a budget, I feel we don't need to be that cheap.

A half hour before the train arrives, we sit in the train station's tea room calmly sipping tea and waiting for the train. The train arrives ten minutes early. This will be the only train that arrives early during our entire time in India. Hundreds of squatting people cover every square inch of the platform. They rise in unison as the train pulls into the station. We realize we're trapped in the tea room. This mass of humanity will be getting on the train with us to Agra!

All four of us are giants compared to the Indians, so we manage to push our way to the train. Pete and Will lift Deide and me through the train window followed by our backpacks. Pete and Will push through the train door. There are no empty seats. People are standing in the aisles and squatting between the cars.

We finally work our way to a sleeper car, where we have no right to be, and keep yelling "tourist" at the conductor, who eventually lets us stay. In the end, for a "slight consideration," he finds each couple a berth to share. We lie head to toe enduring the long, hot trip.

At each stop women with baskets of yellow mangos atop their heads approach our train window. It's so hot we buy several. It's my first taste of a mango which will start my life-long love affair with the fruit. Its sweet juice drips all over us leaving us sticky, but satisfied. The Mangoes sustain us through the ten-hour trip. Lesson learned. No more third-class unreserved train tickets in India.

The Ganges shimmered with life and death a sacred paradox. Watching the pyres burn, I felt the weight of humanity's fleeting existence.

CHAPTER 23

Taj Mahal: a Labor of Love

"How did you sleep?" Pete asks as we leave the sleeper train in Agra.

"Good, considering you were kicking me all night," I say. Deide, Will, Pete, and I take a tonga (horse-cart) to our hotel.

"What a train trip! I'm so tired," Will says, "Let's just go to our hotel." Sitting in the tonga we see the Taj Mahal rising out of the early morning mist. It is translucent and gleaming, dominating the cityscape. Sitting along the muddy Yamuna River, it is perfection in white marble.

"Wow, whoever thought a building could be so beautiful?" Deide comments. "Let's dump our packs at the hotel and head to the Taj."

"Maybe we'll still catch it in this light," Will adds.

At the hotel we hire a van with a guide. Aja Sharma in perfect English tells us, "The Taj is a tomb built by Shah Jahan in 1632 for his favorite wife, Mumtaz Mahal."

We walk through a two-story sandstone entrance. Inscribed on its archway is the Daybreak passage from the Qur'an in Arabic inviting the faithful to enter paradise.

"We need to put on foot coverings to enter the mausoleum."

"This is paradise!" I say as we walk through the entrance. In front of us is the long rectangular reflecting pool and its surrounding symmetrical gardens that end at the magnificent tomb.

Aja tells us, "Look the four octagonal minarets at each corner of the tomb. They seem to lift the building skyward."

We walk the complex in wonder. The intricate marble inlay glimmers with lapis, jade, and turquoise, a testament to the artisans' skills.

"There are no humans or animals depicted on the building for this is not allowed in the Muslim faith," Aja says. "The columns and walls are decorated only in geometrical designs."

Aja whistles loudly. It reverberates several times in the chamber. "You can hear how good the sound is under the dome. Don't share a secret here."

Aja continues, "Mumtaz Mahal and Shah Jahan are supposedly buried in rectangular tombs hidden by these finely carved marble screens, but their real burial places are one floor below us behind locked gates."

On each side of the Taj are mosques. We go to the mosque on the left. Both mosques are made of red Siri sandstone. Their color and texture make a pleasing contrast to the white marble of the Taj.

"These twin mosques on either side of the Taj add to the harmony and balance of the building," Aja says.

Will positions himself behind the red arch and snaps a photo. This unusual perspective and the contrasting red of the arch perfectly frame the Taj.

Aja tells us, "Shah Jahan planned to build his own mausoleum across the river out of black marble, but before he could do it his son,

Aurangzeb, imprisoned him. The son claimed the romantic Mughal was wasting too much time and money on the frivolous building." The Taj now draws seven to eight million tourists a year. Bringing in lots of money for India.

"Imagine how much Shah Jahan must have loved Mumtaz. How romantic,," I say to Pete. "Shah Jahan had three wives but he gave her the title of 'chosen one of the palace'. She died quite young giving birth to her fourteenth child."

"I can't afford to build you a marble tomb, and we better not have fourteen children, but how about a picture?" Pete snaps a photo of me by the reflecting pool. Years later, it would hang in our home, a reminder of love's enduring beauty. Standing before the Taj, I marvel at its symmetry—a love letter carved in marble, whispering devotion across centuries.

From the Taj we go to the Red Fort. Two thirds of the fort are still a military base in 1974. Aja tells us "Here the son, Aurangzab, the third child of Shah and Mumtaz Jahan, imprisoned his father. From his latticed window Shah Jahan could see the Taj. But as his distance sight failed, he had a mirror inset in a column so he could focus on the reflection using his near vision which allowed him to continue to gaze at the Taj and dream of his true love."

"What a sad ending to a great love story," Deide exclaims.

"How wonderful to be loved with such devotion—both romantic and heartbreaking," I add.

An hour southwest of Agra is Fatepur Sikri, the abandoned city. We stop in the village of Sikri to eat lunch ordering what Aja has recommended, Paranthas (chapatis stuffed with carrots or potatoes) topped with hot chutney, a specialty of Agra. It's delicious and cheap.

"I will tell you the story of how Fatepur Sikri came to be. It was built by the great Mughal emperor who ruled for close to fifty years, Akbar. He picked the site for his capitol because a Sufi Saint in Sikri predicted his heir would be born in this village, and he was."

"Akbar himself managed the building of the city which wisely has building styles from Persia and India. Akbar realized he must win over the newly conquered Hindus by celebrating their culture as well as his own. He pulled together all the best craftsmen in the empire to do the work. The city is magnificent, but after only thirteen years, he abandoned it," Ari tells us.

"Why? What happened?" Pete asks.

Ari explains, "Maybe Akbar should have asked an engineer instead of a Sufi Saint where to put his city because the lake that supplied water to the city dried up. He only lived in the city from 1572-1585 after which he moved the capital to Delhi both for practical and strategic reasons."

Through the dusty haze we see the abandoned city of Fatepur Sikri rising up from the flat plain atop a rocky ridge. We drive the van right up to the grandest of nine gates into the city, Buland Darwasa. As we climb the stairs, Ari tells us about the gate, "This gate was built after Jama Masjid, the largest mosque in Mughal India at the time. It forms the southern backside of the mosque and is one hundred and seventy-seven feet high. It was built to commemorate Akbar's victory over Gujarat in 1575. It is called Door of Victory or High Gate."

We look up at a massive rectangular block of red sandstone with an arched opening decorated with cut mosaics of flowers and geometric patterns.

"The inscriptions above on the archways are from Mariam-uz-Zamari, Akbar's favorite wife. It's a good message to those entering a holy mosque, but her luxurious palace contradicts her words," Ari translates the Hindi for us.

'The world is a bridge, pass over it,

but build no houses on it.

He who hopes for an hour

may hope for eternity.

The world endures but an hour.

Spend it in prayer,

for the rest is unseen.'

As we walk through the gateway multiple halls and rooms branch off to left and right. The gateway opens up to a courtyard as big as two football fields and almost as wide. On three sides are yellow and red sandstone arcades shaded by eaves held up by intricately carved stone brackets. Semi-open, elevated dome-shaped pavilions called Chhatris sit atop the arcades. Ari tells us, "These Chhatris are a distinctive mark of Mughal architecture. Akbar won over his diverse kingdom by being tolerant of other religions and honoring their architectures."

In the middle of the arcades is a square pool. Ari says, "This pool is for cleansing before worship and is called an ablution tank."

"Look at the far end of the courtyard. The white marble building is the tomb of Salim Chishti, a Sufi saint. Salim was Akbar's spiritual leader. Although Akbar was illiterate, he built a school for studying Islam in the tomb. He housed poets, writers, artists, and musicians from all parts of the world with many different religious beliefs in the city. Jesuit priests brought pictures of Jesus to the Shah," Ari adds.

Ari points to the western end of the courtyard facing Mecca. "This is the prayer hall. The space is intricately decorated with bands of Persian and Arabic calligraphy, glazed tile work, geometric marble inlays, and polychrome floral paintings. Prayer niches called mihrabs are spaced along the wall facing Mecca."

"Were women allowed to walk around the city freely?" I ask.

"Oh no, memsahib, but the women of high birth and their female relatives lived luxuriously tucked away in inner apartments called zenanas or harems where no men except the important men of the kingdom were allowed to enter. The servants, doctors, artisans, musicians, and guards were all females. Over five thousand women lived in Akbar's zenanas."

"Akbar built three palaces in the city for his Muslim, Hindu, and Christian wives. The marriages cemented alliances within his kingdom. His favorite wife was Jodha Bai, a Hindu also known as Mariam-uz-Zamari. I will take you to her palace, Jodha Bai Mahal. It is the largest and most luxurious of all the palaces," Ari continues.

"How do they know what these harems looked like if no one could enter?" Will asks.

"They don't know, but it's rumored that Johha Bai's palace had valuable carpets, a mural of flying angels, running water, mirrors, and paintings," Ari tells us.

"This courtyard looks like it was once quite lovely. Look at the marble, fountains and walkways. Too bad, there's no plants or water now," Deide comments.

We walk around visiting several other pavilions. Ari points out the Pachisi Court.

"Dignitaries played a chess-like game on the different colored stone squares. Live servants dressed in symbol-like costumes moved around the board as commanded," Ari explains.

We go to Akbar's private quarters and enter a room with latticed walls throughout. Ari tells us, "This is where Akbar played hide and seek with his harem women."

Nearby there's a small grassed courtyard with a rectangular-shaped stone the size of an ice chest in the middle. "What do you think this is?" Ari asks.

"Another game? Possibly of strength?" I answer.

"It looks more sinister to me," Pete says. "Is it an executioner's block?"

"Yes, Pete you're right. It is an executioner's block and the executioner was Akbar's favorite elephant. On the emperor's command the elephant would step on a disobedient subject's head placed on the stone, squashing it like a pumpkin. Although Akbar was considered a kind ruler, he could be ruthless at times," Ari explains. We leave the complex and head for our van.

Fatepur Sikri is resplendent in the sunset as we drive away. This city, a masterpiece of Indo-Islamic architecture, was built mainly to afford leisure and luxury to an elite few, while the majority of the population lived in poverty. All its nameless builders, architects, masons, and artists are forgotten, but India can be proud of this magnificent past civilization.

We give Ari a well-deserved twenty-rupee tip for the tour. We're not surprised when he tells us he has a doctoral degree in history.

It's hard to imagine now, how short our stay in Agra was because in my mind it occupies such a large chunk of how I remember and feel about India. The romance and mystery of Shah Jahan and his beloved Mumtaz. The glory and financial folly of a tomb and poorly planned city. India is a country of unexplainable emotional and mysterious chaos.

CHAPTER 24

From the Desert to the Metropolis

Pete and I leave Agra for Jaipur on the bus. Deide and Will stay in Agra, but we plan to meet up again in Delhi. Just outside Agra we see an amazing sight. A row of peacocks is perched on an adobe wall with their iridescent tail trains hanging down on one side. They turn as we pass, mewing a sweet goodbye.

"Don't peacocks usually screech? These peacocks are meowing like cats," I notice.

An Indian traveler in the seat in front of me overhears my question and replies, "Peacocks screech when they're alarmed, but mew to attract the attention of other peacocks. I guess they like our bus. It's a good sign to see peacocks at the beginning of a journey because they're known as protectors."

"You can never have enough protection," I say as I finger my St. Christopher medal on my pack.

The peacocks faded into the distance as we cross into Rajasthan their iridescent beauty lingering like a blessing on the road ahead. We are now in the "land of kings". Although it's known as the desert state, the landscape is dotted by mustard fields of citron, and chartreuse green. Gray bodies of renegade cattle pop up amid the seas of green. Along the creeks grow neem, peepal, and mango trees.

We pass a wall completely covered by drying cow pies stuck on it. In the middle of each one is the handprint of the person who put it there. Once the cow dung dries it's used as fuel for adobe ovens.

"Look at that wall. Doesn't it remind you of the plaster hands we made for our moms on Mother's Day?" I ask.

"Yeh, it does. I bet every house in the US has these hanging in their living rooms only ours are made of plaster," Pete laughs, "and smell better."

Fifty-five miles west of Agra we stop at the Bharatpur Bird Sanctuary where a metal interpretive sign in English gives us the history of the sanctuary. It was created from the Maharaja of Bharatpur's hunting grounds. Maharaja Suraj May diverted a nearby river to make the wetland. The seven-foot warrior was a secular ruler who defeated the Mughals and the British Army, forming the only independent state of India during British rule. As we travel through the park, we see owls, storks, herons and small Chital deer.

Midway through the sanctuary we stop in a clearing where there's a large concrete sign with the Maharaja's hunts on it. It tells how many tigers, leopards and birds were killed, who was there, and when.

"This is an incredible number of animals killed," I say pointing to the sign. "In 1938 Lord Linlithgow, the British Viceroy of India, killed 4,273 mallards, and seven tigers."

"It wasn't a fair hunt, but a slaughter. The sahibs sat on elephants while peasants beat the forest rounding up game to go toward the hunters. Ten or more guns would shoot simultaneously killing all the animals. It was a virtual shooting gallery, not sport." Pete shudders with distaste.

Now the animals are protected. It was renamed Keoladeo National Park in 1982 and designated a World Heritage site in 1985. Today it is a vast bird sanctuary which brings one hundred thousand bird-watchers a year to India.

The farther we travel, the drier it gets. There are camels, tents, brown earth, and eye-popping saris of bright yellows, reds and purples. Looking like miniature Brahmin cows, zebus wander through the streets. It's the only place in India where we see these small cows.

Pinkish winding walls lead into Jaipur. In the fading light a soft pastel glow from the sandstone buildings shines over the city.

In the center of town, we see the Hawa Mahal or Hall of Winds. This five-story, fifty-foot-high structure rises up over the street. The pyramid-shaped building is a honeycomb of 953 small, latticed windows in the back of the palace. Each window is a miniature work of art with grills, finials, and domes of pink sandstone.

Once bejeweled women in purdah were hidden from the world behind these intricate windows. Through the windows they were able to watch everyday life and festivals being celebrated in the street below without being seen. The small open windows allowed cool breezes to pass into the palace's courtyard.

"I can't imagine sitting in front of a window my whole life watching the world pass by," I say.

"It sounds lavish and luxurious to me. Much better than fighting on a battlefield or plowing a field," Pete responds.

Surprisingly, the front of the palace is only three stories high and plain-looking compared to its ornate and unique back. The Imperial Door opens to a courtyard with fountains and is surrounded on three sides by arched cubicles. Colored glass windows create small rainbows as they shine on the water in the fountains. We are entranced by this ancient "pink palace",

Our pink palace for the night will be the Ashoka Hotel. It seems the entire city has a love affair with the color pink with most buildings sporting the pastel color. As we go out the nest morning a shower of red dye cascades over us. The thrower smiles back at us saying, "Wishing you a very colorful and happy Holi!" and then disappears down the crowded street.

"What!" Pete shouts.

"Look. Everyone's throwing red dye at each other. Retreat to the hotel. We need to find out find out what Holi's all about" I scream over the crowd. We duck back in the Ashoka's lobby where the clerk says "Today is the second day of Holi or the Dhulandi Festival. It is a most joyous time because the monsoons have come. It celebrates Prahalada's victory over the Demon King. It is a festival of colors, love, and spring. We let go of our inhibitions and smear bright colors on each other."

"Is there a place where we can watch, but not have colored dye and water thrown on us," Pete asks. "

"Yes, sir. The dyes can ruin clothes so I will take you to the top of the building where you can watch from the roof," We follow the clerk up a narrow set of stairs to the roof.

Below us in the street, women with arms heavy with gold bangles wave hands covered with intricate henna designs. They swirl and sparkle in boldly colored saris interwoven with silver and gold threads. Mirrors in their bodices, hems, and head scarves reflect the light. Songs rise up from the street and drums beat. Groups of women laugh and sing from large swings hanging from the lampposts and trees.

"Let's get out there and join the party," I say to Pete

"I don't think we're dressed well enough to join. Besides, we don't have enough changes of clothes to ruin even one," Pete answers as he eyes the street wearily.

"Come on. It looks fun. We can always buy more clothes. I've always wanted a sari and you can get a Nehru shirt."

"No, I'm not going down into that chaos. It looks dangerous. Don't pout like you're Cinderella and can't go to the ball. I promise there will be more parties in our future."

"OK, we'll stay safely above it all, but you owe me a wild party." We continue to watch this noisy, joyful celebration from the rooftop.

The craziness of Holi ends the following day so we go to explore the city.

"Did you know that Jackie Kennedy visited Jaipur in 1962 and bought precious and semi-precious stones. The late maharani Gayatri Devi brought her to the Gem Palace to find jewelry fit for a rani (queen). Jaipur is considered one of the best places in the world to find high quality stones," I tell Pete. "Jaipur is known for its first-class craftsmanship and large variety of gemstones. Royalty around the world wear their jewelry. The art of cutting, polishing and making jewelry has been passed down for centuries through their families."

"What are we waiting for? Let's buy some stones. Do you know anything about them?" Pete asks.

"Well, a precious stone has faults, a semi-precious stone doesn't. Precious stones are harder and don't lose their glitter easily," I say.

"So, what are the precious stones; diamonds and rubies?" Pete asks.

"Yes, but also emeralds, and sapphires. They're expensive and rare so we won't be looking for them considering our budget," I tell him.

"Personally, I like semi-precious stones better because of all their different colors and warmth. My favorites are pearls, moonstones, amethysts, opals, garnets, and my birthstone, aquamarine."

"Can you tell what's a good semi-precious stone and which isn't?" he asks.

"A good semi-precious stone should be eye-clear without any defects. They're made from pieces of mineral crystals, rocks, and organic materials. They're also soft and scratch easily. I talk a good game, but we'll be lucky if we don't end up with some plastic fakes," I tell him.

"That'll be good enough," Pete says. "Look we've found the gem street, Johari Bazaar."

In front of us is a long street lined by columned stalls with house facades hanging above them. Outside the shops are sidewalk vendors displaying strings of purple-red garnets, clear moonstones and blue lapis lazuli. A merchant ushers us into his store where we sit on carpets in front of a low table covered in velvet.

"What are you looking for my friends," the man asks.

"Aquamarines, opals, moonstones, and purple amethysts," I answer.

He disappears to the back of the store and then reappears holding four carefully folded paper packets. He opens each pack slowly. Five aquamarine gems spill out on the black velvet. The stones are the color of seawater; transparent with a blue-green hue. Pete holds each of the aquamarines between his thumb and middle finger and peers at it as if he knows what he's doing.

The merchant opens the other packets and the gems shine brightly on the dark cloth. The opals sparkle with blue, white, and purple specks. The moonstones shimmer as their lustrous pearl-colored stripes move across the clear surfaces of the stones. Amethysts colored deep rich purple, vibrant violet, and pinkish violet glitter beside the other gems on the table. Peter continues to hold up the stones and examine them carefully. I pick out three of the aquamarines, four moonstones, a large purple amethyst, and all the opals.

"How much?" I ask.

"Four hundred rupees for all. Very good stones. Good price," the jeweler says.

"Two hundred," Pete counters.

"Two hundred forty," the seller replies.

"Done," Pete says. The shopkeeper carefully wraps the gems and puts them in a velvet string bag. The stones have cost us twenty US dollars.

"What a bargain," I say.

"Only if they aren't plastic," Pete replies.

The following day we take a motor rickshaw to the Amber Fort and Palace twelve miles outside Jaipur. A mile from the fort is a village where you can hire an elephant to take you to the fort's main gate like Indian royalty. It's too expensive for us.

Recently the Maharaja of Jaipur gave the Amber Fort and Palace to the city. The Maharaja has allowed the complex to deteriorate. The ramparts are crumbling, the mosaics are missing large chunks, and the chambers are black with mold. Monkeys roam the buildings. We can only imagine its former grandeur.

The fort is a huge sandstone and marble structure made of four courtyards each with its own gate. It dominates the hilltop overlooking Maota Lake and the plains of India. Jaigarth Fort on the Hill of Eagles sits next to and above the Amber Fort. Tunnels connect the two forts so royals could escape to the higher fort in case of attack. Parts of the tunnel have caved in so we're unable to go underground to the other fort.

As we walk through the fort, an Indian man in western clothes befriends us telling us about each area in the fort. "This complex was for the soldiers and their horses. Here is where the Rajput Mahrajas, their families, and attendants lived. In this court are the two audience halls, one private and one public."

We walk to the next level and see a small temple. "This elegant marble temple is called Sila Devi. This is where family worshipped and made animal sacrifices during the festival days of Navrathri" We enter the mirror palace where mirror pieces cover the ceiling and walls. Even though many mirrors are broken and missing, the building glitters with reflected light.

We enter a courtyard with water channels running through it. Our Indian friend says, "This was the Sukh Mahl (the hall of pleasure). Imagine it with silk curtains, erotic art and pillows. It was the coolest place in the palace. Near it is the zenana courtyard where all the women lived. It is honeycombed with living rooms."

"Thank you for telling us about the fort. What a treasure this fort and palace are for India!" I say. "I hope they will restore it in the future. India's past cultures are civilized and lavish."

I'm glad you appreciate it. So many people from the West know nothing about it," he replies. We share a ride back to Jaipur with him and part after enjoying a cup of chai together.

That night we buy our train tickets for Delhi. When we arrive at the station, Pete goes to the ticket office and hands the agent our tickets to Delhi. "We have a ticket for New Delhi today."

"No train," the agent says.

"What do you mean no train?" Pete says loudly.

"No train," he repeats.

"You have to have a train. I bought a ticket for it," Pete shouts.

"No train. Train left after midnight," he says wagging his head nervously.

I tug on Pete's arm. "He's right, Pete. Our train was for half past midnight on Tuesday not noon on Tuesday. They use the international twenty-four-hour clock, not AM and PM. It's our mistake. The train left twelve hours ago."

Pete turns and looks at me and then begins laughing, "That's two bucks down the drain. Do you think I can ask for a refund?"

I laugh back at him and say, "No refund. Don't even ask."

Pete turns to the clerk and raises his hands, palms together and says, "Namaste, kshama maangana (sorry)," There are no trains leaving when we want to go, so we end up taking the earliest bus to Delhi.

It's a hair-raising trip. Although the road is paved, every kind of transportation is using it. There are carts, elephants, horses, carriages, motor rickshaws, and cars. We stop, start, and roar around all the slower travelers; spewing diesel fumes and missing oncoming traffic by inches.

As the bus pulls into Connaught Place at the center of Delhi, we're exhausted. A four-hour trip has taken six hours. Leaving the bus, we're thrown into the big city cacophony of hawkers shouting, cars honking, and buses backfiring. Rajasthan's serene mustard fields gives way to Delhi's chaotic bustle. Each a distinct facet of India's vibrant identity.

After a good night's sleep, we're ready to explore Delhi's rich past. The city has been destroyed and rebuilt many times with each new conquering kingdom. In our search for the Afghanistan Embassy where we need to get a visa, we look forward to seeing the buildings, monuments, and temples from Delhi's glorious past.

Pete negotiates with the driver of the motorcycle rickshaw which is hard because we have an address, but no idea how far it is from Connaught Place. Pete hands the driver a paper with the address of the Embassy in Hindi.

"Do you know where this is? Afghanistan Embassy. Can you take us? How much?" Pete asks.

"Yes. I know. Very far. Ten rupees," the driver answers. Pete gives the driver ten rupees and we take off.

The driver drops us at an empty lot with a sign that says, "Site of the new Embassy of Afghanistan." We get out, see the sign, and turn to see our motor rickshaw speeding away. We're stuck. We walk to a well-traveled street and flag down another motor rickshaw. Pete waits until we're at the gates of the real Afghanistan Embassy and then pays the driver. We are slowly becoming more travel-savvy with every ride. Don't pay until you get to where you're going.

Our Afghan Visa is processed in one day which is good news because we had booked a Meditation Course in Dalhousie when we were in Kathmandu and now, we only have three days to get there. We'll need to travel back across the country and head north to be there on time.

We have just enough time for a last dinner with Deide and Will. We walk from our hostel down the ring of colonnaded Georgian-style buildings and enter the expensive Embassy Restaurant. Our waiter comes quickly to our table.

"Namaste, I am Madan Singh and I will be your waiter tonight. What would you like to begin with?" he asks.

"What do you recommend? We want to try as many different kinds of Indian food as we can," Deedi answers.

"I recommend beginning with pakoras and samosas. Both are fried fritters filled with spiced vegetables and served with Indian mint sauce, mango chutney, and yogurt. Every state in India has its own unique cuisine. Would you like dishes from different parts of India?" Madan asks.

"Yes. It sounds like an interesting food adventure. What do think, guys?" Pete replies. We all second the idea.

"First, I suggest Delhi's specialty, buttered chicken, murgh makhana. It's a curry in a spiced creamy, tomato sauce. Next, you should try Rogan Josh from Kashmir. It's braised lamb chunks in a gravy made of yogurt, onion, garlic, ginger, and Kashmiri chiles. From Southern India masala dosa is excellent. It's a rice crepe filled with potato curry and condiments of hot sambar and coconut chutney. Lastly one cannot leave India without eating tandoori chicken from the Punjab. The chicken is marinated and cooked to tenderness in a cylindrical clay oven.

"With your pakoras and samosas, would you like a cool, refreshing mango lassi? It's a sweet yogurt drink," Madan asks.

"We're in your hands, Madan, bring the lassis and fritters," I say. Our table fills with food. We pass around the dishes, sopping up the sauces with chapatis (Indian bread).

For dessert we split a mango sundae and a plate of gulab jamun. These are round donuts made of flour flavored with saffron and green cardamon and then dipped in a sweet, thick syrup made with rose water. This meal ranging from spirited to honeyed has left us pleasingly satisfied. How well this spicey, zesty food fits India with its wide range of sounds, sights, and people. Tomorrow we'll head into the northern foothills of Dalhousie to uncover more of India's vast variety.

CHAPTER 25

Sitting Still for a While

In July we'd heard about the Dalhousie meditation course in Kathmandu, and signed up for $50. When we checked our mail in Delhi there was a handwritten postcard from the organizers telling us to meet at the Grand View Hotel for the ten-day course running from August 1- August 11.

Both Pete and I went to college during the "Age of Aquarius" where exploring your inner self through drugs, meditation, or encounter groups was normal. I never did it because I was afraid of the darkness it might unearth. I'm feeling tentative and scared as we climb off the bus in Dalhousie.

Traditionally dressed Indians had filled our bus for the trip here so it's a shock when we reach the town and see young Westerners walking the streets in bell-bottom jeans, headbands, long hair, tie-dye dashikis, backpacks, and love beads. Obviously, Vipassana Meditation is now the latest western fad and we're part of it.

It's August first. We've slammed through much of India in two-plus weeks to be here on time. We're directed to a hotel on the top of a hill. Dalhousie's colonial charm fades as we enter the meditation hall—a stark, silent space that mirrored the stillness we were meant to find within.

"Is this what you expected," Pete asks me.

"I thought it might be nicer, but I've never done this before so I really didn't know," I answer and add, "But it'll be nice to be still in one place for a while," Soon, we realize just how still we'll be. Several surprises await us at the retreat's orientation.

We are led into a large dining room with folded sheets spread out in rows on the floor. At the front of the room is a carpet with a microphone in front of it.

A sari-clad woman welcomes us and explains the meditation retreat parameters, "Namaste my friends. There are several rules to follow during the retreat. You will be housed in male and female dorms. Couples will be separated by gender." Then she adds, "To get the full benefits from the meditation there is no talking. Smoking is not allowed on the grounds. You must limit your sleep to delve deeply into your unconscious self. Meals will be vegetarian and no alcohol or soft drinks are allowed."

With that she dismisses us, "Go to your assigned rooms and unpack. We'll meet back here in the main assembly area in forty-five minutes."

I look around the room. Most of the group are ordinary-looking people of different ages, ethnicities, and professions.

Pete and I look at each other wide-eyed. "Are we going to be able to do this for ten days?" I whisper to Pete.

"I had no clue what this course would be like. I'm gonna miss my smokes," Pete says, "I did a marijuana-fueled, reveal-all-secrets encounter group at UCSB (University of California at Santa Barbara) and hated it."

"No talking will be hard for me," I admit.

I have five Indian college students from Sacred Heart College as roommates. Although I'm only a few years older than them, I'm amazed by how protected and naive they are. They don't date or drink, live with their families, and never go anywhere without male chaperones. How different twenty-year-old Americans are compared to these girls. This course must feel like a silent, pajama party to them.

The first night we go back into the large, bare banquet hall. Goenka, our meditation teacher, sits at the front with his legs crossed under a black and white checkered sarong and wearing a white short-sleeved-collared shirt. His white and gray streaked hair is trimly cut and combed. He smiles calmly at us, sitting on a worn rug. There's no dais. He says, "A teacher should not be made an idol, like a god. He is a teacher. If you want to get any help, you practice what is being taught, that's all."

In 1969 Goenka left Burma to teach Vipassana Meditation. He had been a Burmese businessman suffering from severe migraines, so he meditated with Sayagyi U Ba Khan. His experience was so powerful he gave up his business, left Burma, and came to India to teach. He will later earn worldwide acclaim for his secular teaching of Vipassana Meditation.

This retreat is Goenka's ninety-second class. He rarely teaches English courses, so there are about fifty Western foreigners along with one hundred other students from India.

Our schedule is; 4:30 wake up, two-hour breakfast, lunch, and tea breaks, ten hours of meditation, sometimes in a group together and sometimes in smaller groups.

Travels with Pete

The course rules are too hard for Pete and me to handle. Midway through the ten days, both of us crack under the strict guidelines. Pete sneaks into the town for smokes, coconut cookies, and chocolate bars. Luckily, he doesn't get caught and shares with me. I talk secretly to Pete during breaks. We meet at the far end of the deck that extends along the backside of the hotel. It overlooks the Himalayas where there's a forest with Silver Snow Monkeys with black faces cavorting.

As we watch, Pete makes up monkey dialogue. "Look there's Curly, Larry, and Mo. Hear no evil. See no evil. Speak no evil. Wow, look how high Fosbury jumped, at least ten feet. I think there's some courting going on between Alice and Henry. He brings her a banana, she takes it, and then pushes him away. He must have farted." We laugh behind our hands. It's a good break from silence and instruction.

Meditation always begins with Goenka's deep, resonant chants—the room hums. My husband takes copious notes on Goenka's brand of secular meditation and likes the clean, healthy food. I convinced Pete to take the course so I'm happy he's enjoying it.

"Leave your religion behind," Goenka says, "I am not against conversion. I am for conversion, but not from one organized religion to another, but from misery to happiness, from bondage to liberation."

Goenka brings up small groups in a semi-circle before him during the large group meditations. It's amazing how powerful his presence is. Then, after meditating for a while, he speaks to each of us separately in the small group imparting a bit of wisdom and insight about our experience so far. He seems able to read our minds.

You never know what lurks in our subconscious. During one of the large group sessions, a painful memory comes bubbling up. A vision of my father appears. When my mother drifted into the darkness of mental illness, he was my sole caretaker, and I depended on him for my security and safety. Soon after I started college, he divorced my mom and remarried. I felt abandoned and alone. In the stillness, my father's broken promises surfaced—a weight I had carried unknowingly. With each breath, I let go finding freedom.

This is what Goenka labels a Sankara (an event from our past that causes pain). I had always been afraid of criticizing my dad for fear of losing his love. I had never acknowledged my hidden resentment until now. This buried, painful thought floated in my mind without judgment. I observe it dispassionately— a moment of wisdom and self-knowledge. I think this meditation accomplished more than years of therapy might have. I forgave my father and myself. It's a relief to have the pain gone. I tell Pete what happened and ask him if he's had any revelations. He looks at me and honestly says, "No, my life has been boringly trauma-free."

We buy two Tibetan prayer rugs and vow to continue meditating. Once back on the road, though, all our good intentions to remain mellow don't last. Our calmness disappears as the chaos of India again surrounds us.

CHAPTER 26

◈

Venice of the East: Srinagar, Kashmir

From Dalhousie, we take the Deluxe Bus north to Srinagar in Kashmir. It's a twelve-hour bus ride, and we meet some fascinating passengers; an Indian defense employee, a Malaysian medical student, and a French erotic painter.

"I've been traveling through India visiting temples where there's erotic art. Khajurhao is the best so far, but in Dalhousie I visited the Temple of Vishu and Nagin." Francois says, "In the temple there's an overload of carvings, but also depictions of the sex act which leave nothing to the imagination,"

"The Indian culture is so conservative it's hard to imagine temples full of erotic art," I say.

"Most of these temples were built in the 2nd century when sex was taught as a formal subject complete with pictures of the Kamasutra. Sex at that time was considered pure and a source for new life. Kama (sexual desire) was considered one of the four parts of human goals for life," Francois adds.

"I do lots of literary and biological research on what I'm painting. Here are some of my latest oils," he says, pulling photos of his impressionistic paintings of labium out of his pack.

"Hmmm interesting," Pete comments looking through the photos.

"I'm not sure I would recognize these as female genitals, but there's a sexual swirl to the overall effect," I say.

He invites us to cruise the rivers and canals of France with him at the end of August. It sounds like an interesting trip, but although we keep in touch, we run out of time and end up not going. It will be another what if…moment on our trip.

We're traveling through a continuous canyon up into the mountains. Enroute, there are lots of army troops and military camps.

"What's going on here?" Pete asks our defense department friend.

"Kashmir borders both China and Pakistan. There's always trouble with our enemies. We need to be vigilant and ready," the military man tells us.

Climbing the final pass, we look down into the Kashmir Valley. It spreads out before us like a carpet of green silk; a caldera of plenty. Although the elevation is over five thousand feet, the summer temperature is a moderate 74 degrees Fahrenheit. It's another British Raj summer retreat, but because the Maharajah who ruled this princely state wouldn't allow the British to own property, they built large houseboats on Dahl Lake in the center of Srinagar. These Victorian-style houseboats are the best places to stay in the area.

At the Srinagar bus station, we step out of the air-conditioned bus into the bedlam we've grown used to. Vendors with food, motorcycle rickshaw drivers, beggars, and trinket sellers descend on us, yelling and pleading. On the bulletin board is a note from Deide and Will directing us to the Galatia houseboat on the central canal. We pick a rickshaw, but the other rickshaw drivers aren't happy with our choice and circle, shout, and pull at us.

A policeman comes to our rescue and escorts us to the houseboat. A rickety plank connects the boat to the canal bank. The boat is dirty, and in need of care. We try to hide our disappointment, but Will reads our faces and says, "Cheapest houseboat in town!"

We'll be sharing the boat with the owner, ninety-two-year-old Baba, and a pair of hippies with their nine-year-old son, Jerome. Jerome doesn't know how to read but can roll a joint and change money.

We go to the bow of the boat where mattresses covered with batik cloth are laid out on the planking. Small wooden portholes look out at the garbage-strewn canal. Mosquitoes cause us to close the portholes, and we end up spending a restless, sweaty night.

Waking the next morning, Deedi, Will, Pete, and I go out to shop and eat breakfast. The market is full of finely crafted items. We buy a walnut box with an intricate inlay of semi-precious stones, whimsical papier-mache boxes, several hand-embroidered wool shawls, and a jacket. We spend a hundred US dollars for it all, including the postage to the US.

While shopping we meet Sue and Rod Morris. She works for the US AID agency in Kabul, Afghanistan, and he is doing contracting work at the US Embassy. They invite us to stay with them when we get to Afghanistan. It will prove to be a lifesaving connection.

The next day, Deedi and Will go to Pahalgam to get out of the city and into the Himalayas. We're glad to leave the Galatia. On a restaurant bulletin board we find the Taj Mahal houseboat advertised. It's a luxury houseboat with large rooms surrounding an elegant dining room and deck.

We take a shikara (a flat-keeled boat like a gondola with a paddler in the stern) to get there. The houseboat floats like a relic on Dal Lake, its ornate wooden carvings reflecting the grandeur of a bygone era.

The houseboat is full of other travelers: two Peace Corps guys from Pokhara, Trudy from Bangkok, and Joe from the Terai. We lie on cushioned lounges, drink beer, and eat several varieties of Kashmiri bread (unleavened bread like naan), including a soft donut covered with sesame seeds.

People plunge off the "Swimming Barge," next to us. It's a rectangular, three-storied anchored float with diving boards and ladders. The lake is mud brown. I've seen people urinating and washing clothes from its banks. Although I love to swim, I stick with drinking a cold beer to cool down.

For dinner we feast on gabulu, a giant meatball in a curd and curry sauce and several delicious duck dishes compliments of the Muslim Kashmiri meat cuisine.

Srinagar is truly Venice-like with the Jhelum River, lakes, canals, and surrounding wetlands. The next day we take a shikara to explore. Poking up from the shallow, mucky water, pale lotuses rise from their flat, lily-pad leaves. They stretch out in wide pink, white, and yellow beds before us. Birds of iridescent greens and blues flit about as orange finches, cinnamon-colored sparrows, and the red-throated Kashmir flycatcher serenades us.

From the marshes, we go to the floating vegetable market, where shikaras full of fresh vegetables, purple lavender, and lotus plants line up along the lake's edges.

Vendors call to us, "Try Lotus. Healthy for you. Float in heaven." One hands me green lotus seeds. "MMMM, sweet and crunchy like celery," I say.

Pete tries some lotus stems. "Yummy. It reminds me of water chestnuts."

Another seller hands us a joint saying, "You want smoke pink lotus flowers? Give you enlightenment. You don't worry be happy,"

I quietly warn Pete. "Don't forget about the myth of Odysseus. Visiting the lotus-eaters, he has to drag three of his sailors away and tie them to the ship's mast to keep them from staying forever in the land of the lotus-eaters."

Pete whispers back to me. "We certainly don't want to linger here forever forgetting about home after smoking lotus flowers," Pete refuses the joint.

Sitting on the houseboat the tranquility of the lake mirrors the stillness I had begun to cultivate within—a fragile peace amid a chaotic world.

Before we leave, we visit the Nishar Bagh Mughal Gardens, where twelve descending terraces of water and flowers run down the middle of the garden. It's rectangular in shape with two mountain ranges framing it. The Zabarwan Mountains serve as a backdrop to the garden and the Pir Panjal Mountains across the lake frame it from the Dal Lake side. This oasis of nature was built and preserved to give peace and solitude to counter the noise of men. I sit quietly listening.

It was built in 1633 by Asif Khan, a minister and father-in-law to the Emperor Shah Jahan. Shah Jahan was so jealous of the beauty of this garden which Asif Khan refused to give him, the Shah cut off its water supply and closed the garden. The garden became dry and desolate until Asif Khan's servant disobeyed the Shah and restored water to the garden. When the Shah found out, he approved the loyalty of the servant to his master and gave Asif the water rights forever. What a lovely story of jealousy and envy overcome by loyalty and beauty.

Srinagar is an ancient city that has gone from Hindu/Buddhist rule to Mughals to Afghan tribes and Hindu Dogras to Sikhs from Punjab to Maharaja Hari Singh, who bartered with the British to become one of the princely states in British India.

In 1947, after Indian Independence, Muslim tribes from Pakistan came to the outskirts of Srinagar to reclaim the jewel of Kashmir from the Maharaja. He instantly aligned with India, who sent troop planes to defend the city.

Today, Srinagar is a disputed territory between India, Pakistan, and the People's Republic of China. We visited Srinagar when it was still a visitor's paradise. Little did we know then that conflict and war would be its future tarnishing this idyllic jewel.

CHAPTER 27

Holiest Site of Sikhism enroute to Lahore, Pakistan

We take a different bus south out of Srinagar than we did coming in. It's faster, but also dustier and hotter. In Jammu we'll take another bus to Amritsar. The Jammu Station is squalid with people sleeping all over the station floor. A holy cow picks its way across the floor and then squats landing a cow pie on a man's head. We stumble through bodies to a not-so-good restaurant and hotel.

We're up early heading to the Punjab, the historical state of the Sikhs, and Amritsar which lies on the border of India and Pakistan. We grab the back bench of the bus, but at each stop, more and more people squeeze onto it. We get closer and closer together until finally; the conductor gives up his seat in the front for us to share,

It's dusk when we pass through the walls of the old city. The sunset burns bright red on the horizon, and the Golden Temple's dome sparkles in the distance. One hundred sixty kilograms of 24-karat gold cover it. The bus takes us across the causeway into the temple or in Punjabi, gurdwada. It is the holiest shrine for Sikhs, where visitors can stay and have meals in the temple for free.

Instead of staying in the temple, we pay for a hotel in the city because Pete isn't feeling well and turning a pale yellow. We need our own room. He has been getting sicker and sicker every day since leaving Srinagar. The night and food in dirty Jammu have not helped. During the long bus trip, he's slept most of the way. I worry and start making contingency plans if he gets much sicker. Every morning he seems to be his old self, but by the end of the day he's worse. Pete started the trip weighing one hundred sixty-five pounds. He now weighs one hundred and thirty pounds.

At the hotel, we meet a friendly Sikh with family in the US.

"Please join me for some Kingfisher Beer. It's the best beer in the Punjab," Singha Kaur tells us as he motions us to sit at his table in the hotel's restaurant.

He asks us, "What brings you to this Sikh holy place?"

"We're traveling overland home to the United States and everyone tells us we can't miss seeing this magnificent temple," I say.

"We really don't know much about the Sikhs or the religion," Pete adds.

"I will enlighten you. There are five signs of a Sikh; never-cut hair wound in a turban, a wooden comb, an iron bracelet, all cotton undergarments, and an iron dagger called a kirpan," he tells us. With each item he names, he shows us it pointing to his turban, undershirt, iron bracelet and pulling out his comb and dagger. He continues, "These signs are articles of our faith for both men and women. Women also have kirpans and can wear a turban. Men and Women are equal in the Sikh religion."

"I like that equality," I say.

Pete asks, "Have you ever had to unsheathe your kirpan?"

"I only did it once when I was in New York City and some thugs threatened me. It's only used for self-defense and defending innocents," Singh says. "We don't allow piercings or tattoos, and our motto is "No fear. No hate.""

"No fear. No hate, I like that," Pete says.

"The Sikh religion is the fifth largest in the world and is monotheistic unlike Hinduism. We're known as fierce warriors. Many Sikhs joined the British Army during the Raj. After World War II, Sikhs traveled worldwide." (In the 1980s, trouble broke out between Sikh separatists and government forces. In 1984 Indira Gandhi was assassinated by her Sikh guards. The Indian government killed eight thousand Sikhs in retaliation. The horror of that time still haunts Sikh families around the world.)

"I've heard of how well-respected Sikhs are for their character and hard work," Pete says.

"Amritsar used to be the capital of the Punjab, but in 1947 during Indian independence India was divided into the countries of India and Pakistan. Our Sikh leader, Madani, opposed dividing the country." Singha pauses and takes a long drink of beer, "Our leader was right. The Partition led to violence and bloodshed. As a result, there was a mass movement of Sikhs, Hindus, and Muslims across the new border."

"Did you lose any family members during that time?" I ask.

"Yes, sadly I did. My great uncle was pulled off a train heading to India and stabbed to death by a Muslim mob." Singha lowers his head, and pauses as he composes himself. Our Sikh friend speaks of faith and resilience, traits etched into the temple's walls and his face.

He continues in a more optimistic vein, "Twelve years later in 1961, they relocated the capital of Amritsar farther away from the border and built a new city, Chandigarh, as the capital for the states of Punjab and Haryana to take the place of Amritsar. The great French architect Corbusier designed the new capital, which is considered one of India's most beautiful cities. Too bad you didn't go there," he adds as he says goodnight and heads to his room. (In 1988 Pete would accept a job running a software company in Chandigarh. We will get to know the city well fourteen years after our 1974 trip to India.)

I notice Pete's beer is untouched which makes me worry. He never leaves a beer untouched. He's pushed himself to talk with the Sikh, but maybe he should have taken care of himself or I should have taken care of him.

After a good night's sleep, Pete looks better so we eat a chapati breakfast and leave for the border. Lahore is only thirty-two miles from Amritsar, so we splurge and take a cab.

When we walk down the streets in Lahore, Pakistani men jeer, pinch, bump, and yell loudly at me. Pete begins walking menacingly behind me to stop the touching. The Pakistani men continue to reach for a handful because it's unlikely Pete will shoot or stab them for offending his wife. Finally, Pete shouts, "Stop," loud enough that eyes turn. The Pakistani men giggle. I fume. It's all those "free love" American films made in the eighties which portray American women as "easy." I ball up my fists and grab Pete's hand.

When we first started traveling in the Hindu and Muslim countries, I changed my clothes to respect the country's cultural view of modesty for women in public. I made my jeans into an ankle-length skirt with material sewn between the legs. I also wear a long-sleeved shirt even in the heat so I won't offend the local customs. My wedding ring is visible. I don't wear a hijab (a head covering), but I've had no problems traveling through India. Now in Pakistan I'm continually bothered. We decide to leave Lahore as fast as we can and buy night bus tickets to Peshawar.

We get on a dilapidated bus back to the hostel. As we go to sit down together, the dhoti-clad conductor yells at us, "No, sit! Women front. Men back. No together," Red betel juice drools from the corner of his mouth. Pete obediently sits in the back with the men, and I sit in the front with the women.

At a stop light, I feel a hand pinching my butt through the gap in the side of the bus. I look down and see a man on a motorbike retracting his hand and looking up at me with a wide grin. The bus starts with a jerk, and I yell, "Don't touch me! Leave me alone!"

"Kris, are you OK? What's going on?" Pete roars.

"I'm OK. A man at the stop light reached through the missing plank of the bus and grabbed me," I yell back.

"Come back and sit next to me," he says. Heads turn. The women around me start clucking.

One woman tells me, "This cannot be, missy. You are mistaken. No one touched you."

I glare at her and go back and sit next to Pete. The conductor wags his head back and forth but doesn't stop me.

Pete asks, "How are you?"

"I'm fine. It was just another grab and go, "

That night we head to Peshawar on the Afghanistan border. Is it Lahore or is Pete's sickness affecting how we feel about the trip?

CHAPTER 28

Welcome to Afghanistan!

Afghanistan now is not a place anyone visits. Today it is a scarred land torn by continuing warfare and turmoil, but it was different when Pete and I traveled there in 1974.

It's August 20, 1974. Our overland route from Sydney to San Francisco continues. We'll cross the Afghani border at Peshawar in Northwestern Pakistan and travel through the fabled Khyber Pass.

Peshawar buzzes with chaos—turbaned men in dusty markets, bandoliers across their chests, their wary eyes tracking every move. There are no women anywhere.

In the morning, Pete greets me with yellow eyes and a fever. "Kris, I'm sorry I can't come with you today. You'll have to get the money, and bus tickets to Kabul alone."

"How do you feel? Do you need to go to the hospital? You look rotten and haven't really eaten since Srinigar. I'm worried," I reply.

"Go to the hospital here in Peshawar? Are you kidding? This city feels like the Wild West—uncivilized and backward. I'm gonna sleep it off and be good to go tomorrow. I don't want to spend one more day here." He props himself up on his elbows and looks directly in my eyes. "Will you be OK? I'll be worried about you until you get back.

No buses or walking. Take a cab that the hotel recommends and wear your scarf over your head. Try to get back as quickly as you can."

"I'll be fine. You know I can handle myself. See you later. Get some rest," I say breezily as I close our hotel door behind me.

The weight of Pete's fever presses on me as I venture into Peshawar's tumultuous streets alone. Each glance from the turbaned men like a threat, each step a gamble. I scurry along the street not knowing where I'm going and pulling my head scarf closer over my face. Finding the bus depot, buying our tickets, and getting money is a blur in my memory. I remember stopping often, breathing deeply, and then moving on. As I step back into the hotel's lobby, I collapse in a chair. My doubts far outweighed Pete's confidence in me, but I conquered my fears and did what I needed to do.

The next morning Pete's up before me, smiling and saying he feels fine. I don't believe him, and hope the trip will be easy. In the dark, we board the bus for Kabul. They tie all the luggage on the roof including our packs which makes us nervous not to have them with us.

We are a mixed bunch of travelers, some Afghanis, several backpackers from different countries, and Pakistanis. The couple behind us is from France, Rachel, and Pierre. Rachel is a stunning brunette with mahogany eyes to match. Pierre is a lanky youth with round spectacles. We introduce ourselves and exchange travel stories in English and French. They are students going home to Paris from Goa, India.

We enter the Khyber Pass and follow the two-lane road through a narrow gully lined with manned gun fortifications every quarter mile. Overhead a thin strip of daylight reaches down into the cold, dark canyon. The Khyber Pass has been the nemesis of many an

Welcome to Afghanistan!

invading or retreating army. We can see why. It extends for fifty-three kilometers and is one of the few ways into this landlocked country. It's a perilous gorge.

At the Pakistani border, the driver collects our passports so that Pakistan officials can stamp them with exit stamps, and Afghani officials can give us entrance stamps.

It's hard for us to hand over our documents to a stranger. Our passports have traveled for months against our hearts, safe in our money pouches hanging around our necks and taken out only briefly within our sight at border crossings. This new and different procedure makes us uncomfortable.

We've heard tales of passports getting stolen and not given back, causing costly travel delays. In addition, passports are valuable on the black market. We sigh with relief when the driver returns our papers promptly. A passport is one thing you can never be without. It has become our second skin, our talisman, and our identity.

Ten miles down the road, twelve Afghani army men stop the bus. They motion with their AK47s for us to get off the bus and yell in Pashtun something we can't understand.

We file off the bus and follow the soldiers around a nondescript mud hut. Behind it is a dirty tent with a cardboard table under it. The uniformed man seated at the table motions us to line up. One by one, we hand our passports to him. He thumbs quickly through my passport and then motions me to step aside. Pete is checked and waved on, but he goes and stands next to me and asks quietly, "Is there a problem, sir?"

"No stamp," the officer answers brusquely.

"Let's look again. The stamp must be there," Pete says. "Kris, help me find the stamp. I'm sure the driver got it at the border." We hold the passport between us as we look through its pages.

"Where is it? Where is it?" I say in a panicked voice. "Don't worry, Kris, it has to be here!" Pete answers.

We thumb past Burmese curlicues, Indian characters, and the colorful New Guinea bird of paradise. We flip through it again and a third time. Where is it? It must be there. We have no idea what the entry stamp looks like because the bus driver got the stamp so it is a futile search.

Out of the corner of my eye I see the passengers getting on the bus. I slam my passport on the table in front of the officer.

"Our bus is leaving. I'm sure the stamp is here. Please let us go," I shout.

Pete pushes past me and stands between the officer and me. He notices the man's mouth tighten into a thin, tense line, and his eyes become slits. This is a country where women don't yell at men.

"No stamp," the officer loudly retorts as he turns to me and grabs my passport on the table in front of him.

With one motion, he spits on it and flings it out into the desert. Then he rigidly steps away from the table and turns his back on us. I scurry into the desert to get my passport. Tears roll down my cheeks.

Pete rushes to the front of the building where the bus had been parked. He gets there just in time to see it disappearing down the road in a cloud of diesel gas and dust with our backpacks on its roof.

"Kris, what were you thinking yelling at that guy?" Pete asks.

"I didn't want to miss the bus and I was mad. He didn't bother to even look again. Maybe the bus driver messed up. Besides, we'd already gone through the border. Who were these guys? Why were they stopping us? What are we going to do now?"

"We'll be OK. I don't understand why either, but we need to go back to the border and get that stamp. There are some trucks out in front. Maybe we can pay one of them to take us back to the border," Pete says.

Our backpacks have been our life and home for six months, and we feel strangely orphaned. Will we ever see them again? Will our clothes be parceled out among the residents of Kabul? Will we meet ourselves walking down the street in Pete's Thursday Island T-shirt and my blue flowered mini skirt?

Maybe we can catch up to our bus and save our backpacks. We hire a beaten-up truck to go back to the Pakistani border. Once there, we go to the Afghani office and ask for the entrance stamp. The clerk indignantly shows us the stamp, which has always been in the passport since he stamped it an hour ago.

"That bastard, the stamp was there the whole time. He just wanted to flex his petty power," Pete angrily retorts.

"Welcome to Afghanistan," I say sarcastically.

We race back to the military checkpoint and show the stamp. It's not the spitting officer from before, which is a relief. We get on another bus to Kabul but know our chances of getting to Kabul before our backpacks is slim. Hopping back and forth between borders has cost us an hour and a half.

When we arrive in the main square of Kabul, we see Rachel and Pierre standing alone surrounded by theirs and our packs.

"We thought you might need these, so we waited for you to get here," Pierre tells us.

"Merci beaucoup. Merci beaucoup, Merci beaucoup," I say. How do you thank random strangers enough for such kindness? They have waited an hour for our arrival.

"Let us buy you something to eat and drink," Pete invites. All four of us set off and find a small restaurant on the square where we enjoy gosh-e fil, a sweet pastry and steaming chai. This French couple will forever reserve a special place in our hearts.

Kabul looks like a page from the Old Testament. We have traveled back several centuries on leaving Pakistan and India. This landlocked, raw country has repulsed all conquerors keeping its uniqueness untainted by the modern world and other cultures.

Mud walls surround buildings, and red sandstone mosques with tiled minarets stand out on every corner. Most streets are unpaved and dry. Shrouded women move along the streets sending up small puffs of dirt as they sway through the tented outdoor markets. Bearded-turbaned men ride miniature donkeys through the city, their pointed black shoes dragging in the dust. They are tall, proud men, some with blue eyes who avoid even eye contact, which is a welcome relief after Pakistan.

The next day I call Sue and Rod Morris, the couple we had met vacationing in Kashmir. When they hear that Pete is sick, they send us to a western clinic and insist on us staying with them.

The clinic writes Pete a prescription, and points to the street where a turbaned street - pharmacist sits on a Persian Carpet at the corner. Animal parts, vials of liquid, and herbs in small bottles cover the carpet.

"No! This can't be the pharmacy," I yell. We need real medicine."

Welcome to Afghanistan!

A passerby hears me and says, "Very good medicine all modern." Then the pharmacist hands Pete six vials of clear liquid (tincture of opium) and twenty tetracycline pills. The English on the pill box and vials reassure us.

Rod and Sue live in a suburb of Kabul with walled-in lawns, maids, and modern amenities. It's heaven to be there. They go off to work while I watch Pete sleep. Each day with the medicine and being in a clean, modern home helps Pete recover. I haven't realized how worried I have been until now. I can finally sleep well again. We eat American food from the Embassy commissary and try to repay our hosts with small errands and interesting stories.

After three days, Pete feels much better and since we don't want to impose any longer on Rod and Sue's kindness, we decide to go to the Bamiyan Valley to see the colossal, 6th century Buddhas and this ancient valley.

We wonder what surprises we'll find as we head into Afghanistan's wild countryside.

CHAPTER 29

From the Eyes of the Buddha

After a day-long trip to Bamiyan from Kabul, we eat and stay at a tourist hostel on the edge of town. Pete is not feeling well and stays at the hostel. Several other tourists and I hire a taxi driver to take us to the valley's giant Buddhas' a short distance away.

In the early morning light, the Bamian Valley stretches endlessly, framed by sandstone cliffs and the towering Hindu Kush. The Buddhas loom, silent sentinels to centuries of history. I climb the stairs in the cliff and step out onto the statue.

Standing on the Buddha's head, I see the valley before me. The arch of the dome curves over me, framing this ancient valley where people traveled the Silk Road for centuries, trading and seeking enlightenment. I imagine all the conquerors, scholars, and tourists who have, through the centuries, stood where I'm standing, feeling the immensity of time and history. Genghis Khan left the Buddhas standing; the Taliban did not. What drives men to erase what time itself cannot?

Sandstone cliffs rise on the sides of the Buddhas, and behind the two standing Buddhas are pathways that lead to the caves. Once two-thousand monks meditated and studied here. Fragments of murals of the Buddha, female devotees, and even horses in red, white, and black can still be found in these 5th century caves. Solso, as the locals call the Giant Buddha, was a landmark for all to stop, mediate, and search their souls for truth.

The Buddhas can't share this panoramic view. Muslim believers sliced off the Buddhas faces considering it an insult to their religion. The Giant Buddha's left arm is also missing, so his palm no longer faces forward in a sign of peace. Was this a forewarning of what was to come, I wonder? Soon this valley, and all of Afghanistan, would be embroiled in war.

We descend slowly down the cliff's winding passageway and head to Band-i-Mer and the lakes. Our Muslim driver slaps his shoe on the entrance to the cave staircase as if this Buddha is an affront to decency. This small act of intolerance and defiance may have led to the larger act of destroying the Buddhas completely.

At Band-i-Mer seven lakes step down out of the mountains flowing from one pool to the pool below. They're naturally occurring infinity pools dammed by travertine mineral deposits. We look at the sapphire blue water cascading from one lake to the next and stop at the bottom lake.

The deep blue lake is clear and the August weather is dry and hot so I decide to take a swim. Luckily, I'm wearing my bathing suit under my clothes. I strip down and plunge into the lake. It's so clear I can see the bottom with fish swimming below. The water is refreshingly cool. My swimming attracts a small crowd. A man calls out to me in English.

"Hello, I am Abdul-Azim. I am a teacher in Bamiyan. This is my family," he says as he points to two girls and a boy."

"Hello, I'm also a teacher from the US," I answer.

"You are a good swimmer. I would like my children to swim." The family wades into the water as I demonstrate the Australian crawl

"Both my girls go to school. I want them to go to college," he proudly tells me.

"Good for you. Women are an untapped resource in this country. Thank you for swimming with me," I reply.

When the Russians invaded Afghanistan in 1979, five years after our trip, I hoped my friend's family was safe. I worried more when the Taliban took over the valley. His educated girls surely would have been targets because the acts of the Taliban forbids girls from being educated.

It's still early when we get back to Bamiyan. At our hostel Pete is still recovering. I give him a detailed word picture of the Buddhas and lakes while we have lunch together. I nervously finish his untouched chapati and notice his silence and indifference as I rattle on about the morning. I leave him asleep in our room. Meditating won't work today. I need movement to push out my worry. I grab our Minolta, walk outside, and meet Maureen, a stunning blonde from Britain, who joins me for a walk. We climb over the hill behind our hostel and find a Kuchi (Afghani nomads) encampment of ten large black yurts.

On our way to Bamiyan the bus had stopped when a camel caravan of Kuchis crossed our path. We waited by the side of the bus as the women unfolded mirror inlaid scarves full of silver bangles, necklaces with pendants of exotic stones, and heavily decorated leather belts. Their wild hair streamed down their backs as dark rimmed eyes looked boldly at us. The men were turbaned, with bandoliers full of 303 cartridges and holding tall antique guns. They stood back from all the wheeling and dealing, but later asked for cigarettes and collected money for any pictures we took. The gypsy band scared and excited me. How free and uninhibited they were—wild nomads traveling in an untamed land.

After reading James Mitchener's book, Caravans, I'd developed both respect and fear of nomads. The book is about an American woman who gets captured by a gypsy tribe in Afghanistan, and is finally rescued by a British mercenary. Maureen and I approach the Kuchi tents cautiously.

A swarthy, turbaned man rushes out of the closest tent brandishing a knife and shouting. We hold our ground.

"Hello? We're just tourists looking around!" Maureen sputters. I give a tense smile.

He returns our greeting with one of his own in Pashtun and motions toward the tent flap which he has opened. We follow him inside.

A yurt is a circular tent made of animal skins with a wide-brimmed hat covering it and wooden stakes and poles holding it up.

We walk into the yurt without stooping. Inside, we're surrounded by woolen, oriental rugs covering the walls. A fire is burning in the center. We sit cross-legged on more oriental rugs covering the floor. Nomads and Westerners inspect each other. Several of the children touch our hair. One girl sits close to me and hands me a bracelet. I take off my scarf and give it to her. They give us a cup of steaming tea and sweet flat bread. We have not been captured, tortured or asked to buy anything. We have just been welcomed into a Kuchi home.

Later in the night at the tourist hostel, local police in khakis with pistols at their sides barge into the reception area. The head policeman demands loudly, "All foreigners come with papers!"

Blurry-eyed, all five of us come into the reception area.

"Where women go to nomad camp?" Maureen and I step forward,

"Yes, we went to the encampment and were treated royally," Maureen says in a clipped British no-nonsense way.

I nod in agreement. Broad smiles spread across the policemen's faces. They have come to check that we'd returned safely.

Is Mitchener's book factual? Have Western women been kidnapped here? Should we have been more worried? It's sad when we all start viewing friendliness with fear.

The next day we leave, returning to Kabul where we again stay with Rod and Sue. We pick up more medicine for Pete and he seems to recover, but I'm anxious to get somewhere with better medical facilities.

How strange that years later the names of Kabul, Khandahar, and Herat will appear regularly across US TV screens with news of more explosions and killings. I flash back to these dry adobe cities teeming with life, and feel sad for the proud Afghanis.

CHAPTER 30

Racing across Afghanistan and Iran into Turkey

On September third, we leave in a caravan of buses and trucks from Kabul through Kandahar to Herat. We will make a semi-circle around the mountainous interior of Afghanistan, going first south to Khandahar and then north again and west to Herat. President Daoud Khan is in charge of the country, but tribal militias make traveling long desert stretches dangerous, so a convoy of vehicles drives together.

We drive through uninteresting scrub until we reach Kandahar. Although Kandahar is the second largest city in Afghanistan, the road is dirt and choked with camels, carts, mules, horse carriages, buses, trucks, and the occasional car.

The streets and kiosks are full of men; dark bearded men with mustaches and large white turbans, men in headscarves and embroidered vests, men with multi-colored taqiyahs (raised, rounded skullcaps) and men in billowing white shirts and pants. We're in the founding and spiritual center of the Taliban, conservative Pashtuns.

In much smaller numbers, women walk the streets in white, blue, and black burqas. A burqa is a piece of material covering the entire body from head to toe with a small grill over the face. Like ghosts, they stream along the streets of Kandahar, bumping into things because they have no peripheral vision. Occasionally a gust of wind picks up the burqa revealing a mini skirt and heels.

Once we leave Khandahar, we travel through miles of desolate desert with snow-capped mountains surrounding us. It's September, one of the hottest times of the year, over a hundred degrees Fahrenheit. At dusk, the bus stops at an adobe hut with a few water bottles and rolls on the store counter. Behind two walls are the bathrooms; one for women, one for men. Flies are everywhere. The call to prayer sounds. People unroll their prayer rugs facing Mecca to pray. After they finish, we get back on the bus and arrive late that night in Herat.

The next morning Pete's eyes are yellow again. I've been monitoring his eating for a while, hoping his appetite would come back.

"You don't look very good today, Pete. I noticed last night you barely ate any of the kabobs and rice."

"I feel lousy again, but I think if I just sleep, today, I'll be ready to get back on the bus to Iran tomorrow. Besides, look outside. We're still in the 13th century. I'll take the rest of the medicine we got in Kabul. Don't worry, I'll be fine."

I look outside at the adobe houses, dirt roads and horse carts, and agree, but still worry.

"You're right this isn't much of a place for good medical, but promise me if you get worse, you'll be willing to go to the hospital. I'm going to explore the city. I'll be back soon with more water and bland food. Maybe I'll even find some crackers. Go to sleep."

In the center of the old city is the Citadel of Alexander, or as it is locally known, Qala Iktyaruddin. It was here in 330 BC when

Alexander came through Afghanistan. In 1974, it is a neglected ruin, but the massive adobe walls and turrets have defied time, wars, and conquerors.

I pass a lovely all-blue tiled mosque. I feel like I'm shopping in medieval Europe. Awnings surround the Citadel with tables covered by carpets, beaded head scarves, hunks of meat covered in flies, and large metal bowls full of red, yellow, and brown beans.

The most distinctive difference in Herat from other places is its public transport. Sleek horses with bridles covered in red tassels and bells pull open carts through the streets. I can't resist taking them around the city.

A glass worker sits in his shop, twirling blue glass in a small adobe oven. I watch as he carefully bends and twists the liquid glass into a bracelet. It's lovely. I buy it.

As I walk through the streets of Herat with no head covering, no one touches me. The men are tall, and dignified with blue eyes. The men and women both stare at me, but I think they're just curious. Later I learn this is a normal Afghan custom. They stare at each other as acknowledgment, not hostility.

I look into an alley, and see empty Coke bottles being filled with dark liquid and capped by a bottling machine. Have the cokes we've been drinking this suspect concoction.? Hot tea now becomes our new beverage. Pete is ill. We can't take any more chances with what we drink or eat.

Herat fades into the distance as Iran's modern highways beckon. Arriving in Iran from Afghanistan is like going from the 14th century to the 20th century. We leave dirt, broken water taps, crumbling mud houses, and cheap Afghani dollars to arrive in a clean country with functioning utilities, tidy brick buildings, and expensive rials. It's quite a leap.

Before anyone gets off the bus, the Iranians spray the bus and pass out pills we must take before entering Iran. Hygenic and modern Iran ensures that their primitive neighbors don't pass on anything unclean or unknown.

"We all just ate some pills without knowing what the heck they are! What do you think they were?" I say.

"Maybe if we're lucky they're acid. That'll make this trip more interesting!" a red-haired Irish girl with a four-leaf clover tattooed on her neck says.

"I don't know, but if the pills cure my husband's central Asian crud, I'll come back and kiss these guys," I answer.

The Afghans at the border have again demonstrated their incompetence and neglected to stamp an elderly British woman's passport. Anxiously she pulls at her hair and taps her foot loudly.

Don't say anything! Don't say anything! I say to myself. I think back to our bad experience at the Kyber Pass, and hope it won't happen again. I look at Pete and will him to be silent. No need to add to this poor woman's distress.

The bus driver never hesitates and takes us all back to the Afghanistan side of the border, goes with the woman into the immigration office, and reappears fifteen minutes later with the woman triumphantly waving her passport in the air. We can now leave for Mashhad, Iran, five hours away

Mashhad is a large modern city known as the spiritual capital of Iran. It boasts many of the country's poets, artists, and authors. We easily find a clean, cheap hotel, but Pete is still not well. Few people speak English, but the ones who do want us to know they're not Arabs, but Persians.

"Pete, we need to get you checked out. Your eyes are still yellow, and you're always tired."

"I'm fine. I don't feel bad. These huge bus rides are killing me. I can't sleep on the buses."

"Well, I'm scared. We need to know what's going on. Do you have hepatitis, gallstones, malaria, pancreatitis, or worse, malaria?"

"We went to the clinic in Kabul, and the medicine worked for a while," he says.

"No answers or diagnosis, though, and you're still not well. Let's try a country with a pharmacy in a building, not on the street. OK? Tomorrow, we have a twenty-hour bus ride to Tehran. Let's see if we can find out what's wrong with you at the hospital here. Modern medical facilities have to be better than the cheap clinics in Afghanistan. Besides, we can fly from Tehran's international airport home to the US if you have something serious," I add.

"I'm not cutting our trip short, although not feeling well doesn't make it much fun," Pete responds.

"It's September. Our money is getting low, and we've been traveling for almost a year. You're down to one hundred twenty-eight pounds," I remind him.

"OK, nursing sister Rohver take me to the hospital."

It takes us a while to find the hospital. We get lost several times with signs in a different alphabet. When we finally get into an examination room, the lab work needed can't be processed by the time we plan to leave tomorrow. The doctor examines Peter efficiently and thoroughly, but only speaks broken English.

"Maybe infection? No blood results so I give you some medicine," he says and hands us some tetracycline pills. Tetracycline is a strong antibiotic for infection. Maybe it'll work.

"Do you think his sickness is serious?" I ask

"Don't know. Watch carefully," he advises.

Having no real answers worries me. What if he gets sicker? What if they won't let him on the plane to go home? This part of the world is not easy for a woman to navigate alone, much less with a sick husband.

The next day we take the long bus ride to Tehran. The city has a distinct air of progress and international sophistication with underground malls, women in Western-style clothes, soup, yogurt, sub sandwiches, and an efficient post office. Mohammed Reza Shah Pahlavi is firmly in control, and from our perspective, Iran seems like a progressive country on the rise.

Right before we arrive in Tehran, the city hosted the 7th Asian Games. Over three-thousand athletes from twenty-five countries participated. It was the first Middle East country to host the games. The Aryamehr Sports Complex was built with state-of-the-art technology complete with photo-finish cameras, a synthetic track among other modern sport innovations for the time. No wonder the city looks so sparkling clean and impressive!

Not far from our hotel near the bus station we find several cafes with submarine sandwiches. Pete wolfs down a sandwich and half of mine.

"It looks like you're feeling better, Pete," I comment.

"Yeh, I'm starved. These subs sure are great," he says with his mouth full. Less than a half hour later he's ordering two more subs and eating them.

"Wow! I think you're cured. I haven't seen you eat this much for a month," I add. "And, the best part is they only cost ten rials each (US thirty cents), Let's fill up a sack for our trip through Turkey," Pete answers.

My ex-Australian roommates, Haley and Rhonda, are working in Tehran at the time, but we've lost touch and have no idea they're here. It wasn't until we're back in the US that we learn we were in Tehran at the same time. After six months there, they will have a much different view of progressive Iran and the Shah.

Pete is fully recovered. Although I wait for a relapse, none comes. What cured him? The pill at the border or the modern country? To this day Pete's illness and his instantaneous recovery is an unsolved mystery.

Our next stop is Ezrum, Turkey. It's a twenty-hour bus trip across Iran, but it's on a modern luxury BUS. Dark-haired men with black mustaches, sporting western fedoras, and smelling of heavy, spicy cologne fill the bus. The Shah in his rush to westernize the country has promoted wearing fedoras over the traditional kolah namadi, a wool brimless cylindrical cap with a rounded crown. Although it's not popular with the conservative, traditional community, many of the men are wearing fedoras. The Shah will learn in the near future how his push to Westernize Iran has gone too fast and too far.

On getting to Ezrum all we want is a hot shower. We move along the row of hotels. "Do you have a room? How much? Hot shower?" Pete asks each hotel clerk. They shake their heads and answer in Turkish.

We keep going until we find a clerk who speaks English, "My hotel is best. Fifteen liras."

"Do you have a shower with hot water?" Pete asks as he pantomimes, showering and fanning himself.

"Oh yes, sir. Lots of hot water," he parrots back to Pete.

We throw our packs in the room, and I head to the shower. I lock the door, undress, and turn on the shower which is cold. I take a cold shower. There's a noise behind the bathroom door. I Look and see two sets of feet below it. I drape my towel around me and look through the keyhole. I see another eyeball staring back at me. I push open the door. "What are you doing? You peeping Toms! Get out of here," I yell. The hotel clerk and another man race down the hall.

"Sorry, sorry, sorry," bleats the clerk.

When I get back to our room, Pete says, "Look, the hotel clerk just brought us hot tea and pastries, Kris. Wasn't that nice?"

"No. Those pastries and tea weren't free. The hostel owner is bribing you to ensure you don't kill him for eyeballing me in the shower. A cold shower, that is."

Pete grabs a pastry and says, "Go take another shower. Maybe we can get a free dinner."

"You take a shower this time. See if your skinny butt gets us anything."

Seven hours later, we reluctantly get back on the bus to Istanbul. It's been too many days of buses, and we're road weary, but healthy. We've spent ten months in Asia and will soon be a day away from Europe.

Our trip is ending, and we think about all the places we've missed: Africa, the Middle East, and Southern Europe. How delusional we were to think we could actually see so much of the world in a year. Traveling will never be "out of our system," but we're tired and miss our families, friends, and country.

CHAPTER 31

Where East Meets West

It's night when we arrive in Istanbul. The city shimmers with blue, red, and gold neon lights as we drive through the centuries-old city's walls passing over the ageless Bosphorus. The youth hostel is barracks-like, the communal bathroom barely functional. But its location, steps from Istanbul's heart, makes up for the discomfort.

Our first stop is The Pudding Shop. We've been hearing about this restaurant since we left Australia. It's the information hub for overland travelers. On its bulletin board are places to stay, messages from friends, and tips on traveling either East or West. It's the 70's "Hippie Trail" internet.

Hi, Millie, I'm here in Istanbul at the Sultan Hostel. Come see me. Simo.

"Filling up a bus headed to Isfahan. Sign up. Leaving the Pudding Shop Wednesday at 8 AM."

"Passengers needed for a car trip up through Bulgaria, Yugoslavia, and Hungry to Austria with Ricardo, an Argentinian doctor. Leaving in three days."

Pete reads Ricardo's posting and then asks me, "Should we sign up?"

"Sounds like he's headed in our direction. It would be interesting to go to communist countries, but it means we'll only have three days here in Istanbul," I say.

"I'm tired of buses. Being in a car will be cheaper and a good change. Let's meet this Ricardo and see what we think. We can decide then," Pete answers.

We write our names and where to contact us on the message board. Looking around, we watch as a young woman in a flowery dress with unruly hair leans across a table speaking intently to a mustached man in a leather vest and bell-bottom trousers. We hear the words, "Cyprus" and "invasion." At another table we see a dark-haired man in a suit laugh as he downs a small cup of black, Turkish coffee and looks across at a well-dressed couple who ask him, "What should we see at the Topkapi Palace?" We down our cokes and head out to explore the city.

Three days isn't enough time to explore this fascinating city, but after days of bus travel, we're bursting with energy and enthusiasm. Hagia Sophia Mosque is our first stop. It was built in 537 AD by Byzantine Emperor Justinian when the city was Constantinople, and Christianity was the practiced religion. Emperor Justinian dedicated Hagia Sophia Mosque to the wisdom of God.

Four minarets surround the enormous dome. It's made of Ashlar (precisely cut and fitted stone) and Roman bricks, which give it its reddish hue. Its marble floor looks like the sea with waves of varied-colored ancient marble. Mosaics of Jesus, his mother Mary, and angels fill the nave and entrances. Light from the dome reflects in the interior of the building making the rotunda hover above us. It's humbling knowing centuries before us such magnificent buildings were made. Will our civilization be able to match this awe-inspiring creation?

Nine centuries later, the city fell to the Ottoman Empire, and the cathedral becomes a mosque. Now it's a breathtaking beauty, a monument to Byzantine and Ottoman architecture. When Turkey became a secular country in 1934 under Ataturk, it became a museum, a monument to Turkey's Christian and Muslim past.

In 2020 the President of Turkey, Erdogan, converted the museum back to a mosque causing worldwide controversy. All vestiges of representational art have been covered or plastered over.

Across the square from Hagia Sophia is the Blue Mosque. We've been warned about the dress code. I wear a scarf and a long dress, and Pete wears long pants. We're unprepared for the magnificence the building. Six minarets rise above the Mediterranean surrounded by a grove of palm trees and gardens. Blue light from the azure tiled walls and domes compete with the color of the sea. People come to worship, socialize, and learn in this heavenly environment.

Next, we climb the hill to Topkapi Palace, a 15th-century palace that Ottoman sultans used for four hundred years. "It looks like we're in the high rent neighborhood," I say as we look out at the view from the palace. Below us is the Golden Horn (a natural harbor of the Bosphorus) lined with luxury apartments and mansions.

"Look at this jewelry! It's better than the Crown jewels of England," I say. Of more interest is St. John the Baptist's right arm, a Christian relic traded for by a Muslim sultan in the 15th century. Gilt medal made in Venice surrounds bones with a small square cut in the arm to show the actual bone. The hand points upward in the Byzantine sign of blessing, but Christian believers say it is the gesture John used to point out Christ as the Lamb of God. The unusual mix of East and West, Christian and Muslim, is fascinating.

When we get back to the hostel, there's a message from Ricardo to meet him for lunch at the Pudding Shop. Since arriving in Istanbul, Pete and I have been stuffing ourselves with bread and an orange soup called Tarhana made of tomatoes, green peppers, yogurt, nectarines, dill, milk, and sweet red peppers. Every tiny stall has it, and it's only a few liras. So, although we've been eating all morning, we join Ricardo in the courtyard of the Pudding Shop for lunch. We wave tentatively at a handsome man dressed in a pair of sharply creased khakis and a cashmere sweater. He's sitting alone at a table.

He rises from the table and beckons us with his hand. "Are you Pete and Kris? Welcome, friends. I'm Ricardo." He's a handsome, dark-haired Argentinian with impeccable manners and English.

"I saw you're interested in joining me on my car trip to Austria. We'll be stopping in Bulgaria, Yugoslavia, and Hungary. I've always been curious to see communist countries with my own eyes. Our President Juan Peron died recently, but not before transforming our country from a military dictatorship into a Socialist Republic. He and his dead first wife, Evita, are my heroes. I'm leaving Sunday, September 15th at 6 PM. I rented a car in Paris and must return it in two weeks. It's a bit bashed up, but it still runs well. "

"This sounds like it'll work for us. We're headed for Europe and want to get there before it gets too cold. We plan on being back in the US before Christmas. What will this cost us?" I ask.

"I'll need your help with gas money, and you need to pay for your food and lodgings. I'm starting to run short on money, so if there are more travel expenses, maybe we can share them. Travel companions are what I want. I'm planning on finding two more passengers before Sunday."

"What kind of car is it?" Pete asks

"It's a white Mercedes Benz station wagon with a strip of black paneling along each side. After finishing my medical studies in Buenos Aires, my parents paid for a European trip. I rented this car and have been all over," Ricardo tells us. Pete then recounts our travels.

"We've backpacked overland from Australia. Sometimes we stay in flea-bag hotels and take local, broken-down buses. Other times we flip-flop and stay in luxury hotels and take flights. Being in a Mercedes is an upgrade compared to our last few trips. Count us in," Pete says.

After lunch, we head to the Grand Bazaar. It's the world's oldest and largest covered market, with multiple entrances and streets geared to specific products; carpets, lamps, textiles, clothes, and much more. You probably can find anything from the East or West in its sprawling expanse.

Inside the market, colors, chatter, and music explode around us. People bustle from one kiosk to the next, and merchants call out greetings, hoping to lure us into their shops. Usually, this many people in such a small space would be scary, but there's an overall feeling of well-being. Heavy clove and cinnamon smell fill the air as we walk from one shop to another.

"This is the perfect place to finish up our Christmas shopping," I tell Pete.

"Who's still without a gift? We've sent so many packages home already. Haven't we covered everyone in our families by now?" Pete asks.

"Actually," I say as I flip to the back of my tiny diary where I've kept a running list of all the presents bought. "I haven't found anything for your dad."

As I say this, we pass a shop with all kinds of smoking devices, hookahs, bongs, and pipes. Inside we find a walnut pipe rack with three handsomely carved meerschaum pipes.

"You like? Good price. Meerschaum found here in Turkey. Best material for smoking. Pure for tobacco taste," the vendor tells us.

Pete takes each pipe off the rack separately examining the pearly bowls intricately carved with geometric patterns. It's a bargaining trick he's perfected during the trip.

"Is this too heavy?" Pete asks me. I scowl and nod my head yes.

"Meershaum very light, sir," the vendor says. "Twenty-five lira."

"Twenty," Pete counters. The vendor wags his head in agreement, takes the liras and wraps up the pipes and rack.

The next day we take a ferry down the Bosphorus Strait to the Black Sea. Luxury villas and palaces line the water. Ashore we sip dark coffee and eat baklava in an outdoor cafe that feels more European than Turkish. Russian ships stream by. The Bosphorus Strait, the Sea of Marmara, and the Dardanelles are the only passage between the Black Sea and the Mediterranean. It's a major sea access route for Russia, Bulgaria, Ukraine, and Romania. Since ancient times these bodies of water have been essential for commerce and strategic reasons.

It's our last day in Istanbul, so we buy first-class tickets on a deluxe ferry to the Princes' Islands in the Sea of Marmara, which costs us ten liras. Skinny, green Cypress and pine trees dot the shore, and the clean water sparkles with a brilliant blue in the intense Mediterranean sun. We take a horse-drawn buggy to the beach because no cars are allowed on the islands. We glimpse large mansions just off the road. I go swimming, and we sit under an umbrella sipping raki, a rich aperitif made from grapes and anis. Sitting on this beach is the closest I will come to the French Riviera. It's inexpensive luxury.

We find out about a ship leaving from Izmir, Turkey, to Athens. This trip tempts us to abandon the car trip with Ricardo for the ship headed to Greece.

"Going to Greece and being on the Mediterranean sounds fabulous, but I bet we'd like it so well we'd stay too long and never make it home by Christmas," I say.

"We've committed to Ricardo, but he's a fellow traveler, he'd understand if we change our minds at the last minute," Pete replies.

"Ricardo wanted to leave tonight. I doubt he'd be able to find two more passengers to fill our spots. Bulgaria, Yugoslavia, and Hungry sound interesting to me. I vote we visit Greece another time. Let's get back to the hostel and pack," I say. We will end up visiting Greece much later on our fiftieth wedding anniversary.

At six PM we meet Richardo at the restaurant. He has found two other passengers; Victor, a Malaysian, and John, a Scotsman. We introduce ourselves and then climb in the car. John takes the front seat and Pete, Victor, and I sit in the back. It's late and Pete and John help Ricardo navigate his way out of the city.

We travel over the Galata Bridge across the Bosphorus and head to the Bulgarian border. We arrive at midnight. The border is closed. We sleep, uncomfortably, in the car until it opens in the morning. Behind us the lights of Istanbul look like distant stars. This unique city feels modern and ancient, Christian and Muslim, Eastern and Western—a tantalizing mix of contrasts.

CHAPTER 32

Going Behind the Iron Curtain

Ever since Pete and I were born, the Soviet Union and Communism have been pitted against the United States and Democracy. We're heading into enemy territory; the Soviet-occupied countries of Bulgaria, and Hungry. We'll also be passing through Yugoslavia which is ruled by Tito, a Communist dictator. These countries have been closed off from the Western world ever since the "Cold War" started.

"Ricardo, do you think they'll let us into Bulgaria, a Russian satellite country, with US passports? John's British living in Ibiza, Spain and Victor is Malaysian. Will any of us need a visa?" Pete asks.

"Sorry, Pete. I don't know. We're only transiting the country so I don't think we need visas. My Argentinian passport is good because of our Socialist president. We'll just have to see what happens," he answers.

"Pete, I can't believe we didn't think of this before," I say with panic in my voice. "Why didn't we ask someone or go to our embassy? Do you think border guards will know any English? None of us know Russian or Bulgarian."

"Don't worry. We'll manage somehow. We always do. Just smile a lot and act confused. Look there's a bus over there. I bet it goes back to Istanbul," Pete reassures me.

We drive through the Turkish border. The Turkish border guards barely look at our passports and motion us toward the Bulgarian border two hundred yards down the road.

The Bulgarian border isn't much of a border when you think of it being behind an iron curtain. A guard house and a flimsy white gate with a round red reflector in the middle mark the boundary of Bulgaria. There are no machine guns, cement walls, or brick barricades. It certainly isn't the Berlin Wall. The guard, dressed in a drab, gray uniform with a green, red, and white striped Bulgarian flag patch on his sleeve wordlessly motions for our passports.

As the guard examines our passports, I hold my breath, bracing for rejection. His curt nod felt like a lifeline as he hands our passports back to us.

As he hands Victor's passport back, he says, "You no go." He pointed back to the Turkish border, "Malaysians not allowed."

Grabbing his backpack Victor slams the car door and stomps toward the bus. "Let's get out of here," Ricardo says as he presses down on the gas. Pete lets out a long breath and I turn and look back at the unfriendly guard who stands with his hands on his hips watching us leave. It's only a two-hour drive from the border to Sophia. We're hungry and tired and anxious to get there.

Four-thousand-foot-high mountain massif, Vitosha, coldly greets us. This enormous granite dome dominates the city—a somber giant hanging over it. Bulgaria felt like stepping into a shadow—an iron curtain that chills the warmth of human connections. We go into a bakery where the shelves are almost bare and end up with stale rolls.

As foreigners transiting the country we must stay at the state-run hostel. It's a long, concrete barrack covered with black mold and peeling paint. There's a male and female wing, so John, Ricardo, and Pete head to the men's branch while I navigate the dark, poorly furnished women's dormitory alone.

Several women look me over as I walk in. I smile, but they pull their headscarves over their faces and look away. Have they, too, been taught we're enemies? We warily eye each other. If they talk to the Capitalist, will they be punished I wonder? I feel isolated and scared.

A young, pregnant woman dares to talk to me. She hesitates for a moment, but curiosity overcomes her fear, and she gestures at my pack, laughs and says, "American." We do a variety of hand charades and saying words in English and Bulgarian to each other as I unpack my sleeping bag and toiletries on the cot. Her kindness in this cold, dark place calms my fear, but I wonder if her comrades will report her. Will she suffer consequences for her brave friendliness? To myself, I thank her for her words.

Later I learn from a CIA agent in Sophia in 1974 that I was right. If a Bulgarian talked to foreigners, he or she could be reported to the Soviets and penalized.

After a night's sleep, we head out to see the city center, only a few blocks away. It's mid-morning, but there are no cars on the cobblestone streets and few people.

"How were the men's quarters at the hostel? Did anyone talk to you" I ask.

Ricardo answers, "Basic, ugly, with a concrete floor, a single sink and two urinals."

"There were several surly guys lying on their beds sharing a bottle of vodka. They never even turned to look at us," Pete adds.

"Only a young girl talked to me. Everyone else shunned me. Had it not been for the girl's friendliness I probably wouldn't have slept too well," I tell them.

When we get to the main square, the Alexander Nevsky Cathedral dominates our view with its onion-shaped gold domes. It was built in the late 1800's. Unfortunately, this relic from Eastern Orthodox Christianity is unkept and empty. One can only imagine its former grandeur. Italian marble, Indian Alabaster and Brazilian onyx line the walls and floors. Delicately painted icons, gold chandeliers, and royal thrones surround the nave.

Back on the street, we go several blocks to the Largo, three gigantic, concrete-block, government buildings. The complex is Socialist Classicism at its best. Red banners hang down the length of the buildings. A giant red star on a pole tops the central building, which is the Bulgarian Communist Party House. When we try to enter the building, a soldier with a machine gun stops us. As we walk away, Ricardo raises his arm and shouts, "Bravo, Che Guevara!"

"Whoa, Ricardo, what are you doing?" Pete says, "These guys don't know Che Guevara and don't care if you're a budding Communist. The shouting and raised arm look threatening. These guards are wound tight. Don't get us arrested. If you think the hostel is rough, imagine what the prisons are like!"

"OK, Pete. Calm down. I just wanted to show some solidarity," Ricardo answers.

The nine-story high TZUM department store is to right of the Bulgarian Communist Party House and the Council of Ministers of Bulgaria is to the house's left. A Yellow pavement forms the center with a patch of grass. The whole complex is stark, gray, and melancholy.

Inside the TZUM department store are jars of caviar, Mustang convertibles, mink coats, and French cosmetics. The Bulgarians walk the marble floors with gold chandeliers above them like a heavenly exhibit. No one in this poor communist country can buy anything in the store; it only accepts US dollars. Pete and I find a few US dollars hidden in our money pouches and buy some Ghiradelli chocolate, the only thing we wealthy capitalists can afford.

The guard at the front entrance tells us an interesting urban legend about the building. He is one of the few people we meet who speak English.

"Welcome travelers! Where are you from?" the guard asks. We tell him, and then he starts telling us the history of the building.

"The building was built in 1952 by Kosta Nikoloo, the main architect. His son was killed in Germany by the Gestapo in 1940. Somehow Kosta was able to bring his remains and place them in the foundation of the department store. In Bulgarian folklore, embedding a living person or their shadow in a building strengthens it. This building will stand forever." The guard smiles proudly then asks, "What did you think of our department store? Communist countries have luxuries too like in the US."

"Yes, you certainly do. Have you ever bought anything in the store?" Pete queries.

"No, but I like to look and dream. Only top Communist leaders can buy.," the guard answers. This uniquely Bulgarian belief about bodies in buildings feels cruel to me. We don't bother to question the inequality of the department store. Somehow, we sense the danger of being too open.

In 2001 long after the Soviet-dominated People's Republic of Bulgaria had become a member of the European Union, renovations on the old soviet department store unearthed Kosta's son's bones, making the legend fact.

We also visit the Soviet memorial to the liberating Russian soldiers from WWII. A lone Russian is protecting a Bulgarian mother and child on a pedestal atop a hill surrounded by the park and other statuary murals.

One of the murals is a realistic sculptural relief of charging Soviet soldiers. In 2011 the mural with the soviet soldiers was repainted as American pop-culture heroes; Superman, Joker, Robin, Captain America, Ronald McDonald, Santa Claus, Wolverine, the Mask, and Wonder Woman with the tag "V Kraks s vremeto." In pace with the times. Dissidents continue to paint the statue despite Russian protests.

All four of us climb into the Mercedes the next morning and head to Belgrade, Yugoslavia. We leave the 5th century city of Sophia where civilizations have replaced other civilizations for centuries and hope that Sophia's next civilization won't be as dark as this one.

On the way, we pick up a hitchhiker, Dario, a card-carrying member of the Communist party. He's headed home from college to see his mother. Ricardo ideologically flirts with Communism and grills Dario, who vehemently defends the superiority of Communism over Capitalism.

"In Communism, everyone prospers, not just the rich Capitalists. Everyone's standard of living is high, and we have no poor people. All have jobs," Dario boasts.

"Do you get to choose the job you want?" Ricardo asks.

"The party knows what job is best for you," Dario answers. "All women serve alongside men in the military. Ours is a much freer society for Negros and women."

"Can you practice whatever religion you want in Yugoslavia?" Pete asks.

"Religion is corrupt and brainwashing. We are more scientific and progressive. Religion is not encouraged."

"Aren't all man-made institutions corrupt and brainwashing by nature? No religion feels soulless to me. Considering all the atrocities man has committed, I'd like to think there's a higher power, God," I answer.

Pete adds, "The US is a country of faith, but also one that values science and progress."

"What a good discussion. I would like you to come to my apartment and meet my mother," Dario says.

"We'd love to," I reply.

Thirty minutes later we arrive at a massive apartment complex of gray concrete rectangles. We walk up five floors and arrive at Dario's mother's apartment. There are two rooms with a table, a galley kitchen with propane burners, a small square ice box, a curtained-off sleeping area, and a worn couch and chairs. He introduces us to his mother, who smiles and then motions him to follow her into the bedroom.

When Dario returns, Pete asks, "Is your mom upset about us being here?"

"No, she doesn't like me hitchhiking," he says.

"Our moms are like yours. They don't like us hitchhiking either," I assure him. Ricardo and Pete nod their heads in agreement.

He offers us a cup of tea and a bar of chocolate. We stay an hour and then leave. We appreciate his openness and hospitality, but this tired apartment is not an American, English, or Argentinian definition of a high standard of living. Even Ricardo has to temper his vision of life under Communism.

We arrive in Budapest, "the white city", five hours later. Its name literally means white city in Yugoslavian. The name is from an ancient fortress that was built on a white ridge which even today has strategic importance due to its location at the confluence of the Danube and Sava Rivers. A warm Yugoslavian family in a grand, old house in the city's center rents us a large communal bedroom.

The city is pulsing with cars, people, and products since Tito broke from Russia. His brand of Communism looks more like prosperous market Socialism, not Soviet Communism. Mini-skirts and pop music fill the streets. The oppressive atmosphere of Bulgaria contrasts sharply with this sparkling capital. I want to look up Yani, my Yugoslavian friend from Nepal, but we only spend the night. After seeing her city, I question her diatribes against Tito. Yugoslavia is thriving and alive.

Ricardo, Pete, and I troop off to the Kalemegam Fortress, leaving John back at the house. The fort is built high above the confluence of the Sava and Danube Rivers overlooking Belgrade. There are several scam artists there who try to lure us into betting games. Ricardo loses a few times and then quits playing.

As we walk along the balustrade, two glamorous women head toward us. They are so well-dressed we wonder who they are. The older woman is dressed in a beige, belted suit jacket and matching pencil skirt with a black beret, scarf, gloves, and heels. The younger woman wears a sheepskin coat over an orange turtle-neck sweater tucked into an orange polka-dotted mini skirt, orange tights and mid-calf dark brown leather boots. Her auburn hair falls down her back to her waist.

"Bon jour. Are you Americans?" the young girl asks.

Ricardo answers, "Non! I am Argentinian, but my friends Kris and Pete are Americans. Are you French?"

"Oui. Je suis Yvette et ma mère Madame Brochet," Yvette answers.

"Je suis Ricardo. Je parle français, anglais et bien sûr, espagnol," Ricardo brags.

"Please, may we speak English so I can practice? Why are you in Yugoslavia?" Yvette asks.

"We're traveling through Yugoslavia on our way to Austria and then back to our countries," I answer.

Madame Brochet joins the conversation, "It's so unexpected to meet young, international people in Yugoslavia. I am the wife of the French ambassador. Our daughter is here visiting us from her studies at the Sorbonne in Paris and has been anxious to meet some young people her age. Come with us to the embassy for an apertif."

The French Embassy in Belgrade (then Yugoslavia, now Serbia) is an impressive art deco building of white marble which is a one-minute walk from the fortress. We walk through a grand and elegant reception area with shallow reliefs celebrating French wealth and culture to the ambassador's apartment on the second floor. It is as elegantly and tastefully decorated as the rest of the embassy.

The ambassador joins us, rolling out a two-tier cart full of liquors so exotic that there's not a single one we know. Madame Brochet serves the drinks in small crystal glasses. To their credit this sophisticated family never makes us feel self-conscious about our vagabond appearance. It is an elegant and refined moment of unexpected luxury and conversation. After cocktails, we go with them to a small restaurant where we feast on fresh bread, cheese, and tomatoes.

On returning to the boarding house, I find my pack open, John gone, and most of my jade jewelry gone as well. The Scotsman living in Ibiza has robbed us.

"Goddamn that bastard. He's a fellow traveler robbing other travelers. Who does that?" Pete says as he bangs his fist on the wall.

"If he needed money, we would have given him some, but robbing us and skipping out on Ricardo is really low," I add.

It will be the only time on the trip we get robbed. Two months later we see John on a London street, but when we call his name, he looks over his shoulder and scurries away like a rat into the crowd.

The family that hosted us in their house sends us off after a hearty breakfast of cheese, bread, sweet pastries, and coffee. It's only four hours north to Budapest. We travel in and out of villages crossing fields of Maize and apple orchards. To the West and Northwest we see the Mata Mountain Range and around Budapest are the Buda Hills some of which are within the city limits. We go into the center of the city and find a shabby, inexpensive hotel.

From there we decide to go to the Gellent Baths near our hotel. They're segregated male and female baths. I sit naked with old women in a curved warm mineral bath with Roman columns edging chipped mosaic walls and floors. The water is lovely, but the women are silent and sad. It makes me depressed soaking with these defeated and depressed women.

After the baths, Ricardo, Pete, and I walk the streets sightseeing and looking for a restaurant, Finally, as the day darkens, we settle on a drab pub close to our hostel. Goulash and paprika chicken are the only menu items, so we order goulash and beers. In the far corner, an unshaven old man is nursing a beer. He looks up as we sit down and calls out, "Are you, visitors? Where are you from?"

Ricardo answers, "Argentina."

Pete chimes in, "United States."

"How do you like our city?" The man asks.

"Quieter than we expected," I honestly answer.

"Ah, you should have been here before the Soviets. Then, there was music, dancing, art, and laughter. Now, we just try to survive. In 1956 we fought for our freedom. Stalin is gone. Nikita Khrushchev says it's a new era. Your President Eisenhower and Secretary of State Dulles say that the US will support the liberation of captive people in Communist countries. The Voice of America radio broadcasts tell us that they will help us if we fight to be free. We fight, but no one comes to help us. The Soviets bring in tanks and troops and kill two and a half thousands of us. Two hundred thousand people left as refugees. Thirty thousand went to the US. The US does nothing. My friends and family get killed because of US's empty promises."

Pete and I knew about the Hungarian Revolution. Seeing the anger and sadness coming from this defeated man, we feel shame and embarrassment for our country and its betrayal of the people of Hungary.

"We're so sorry our country never helped you," I say softly.

"It was a mistake," Pete adds. The man slumps his head on the table and sobs.

This sad scenario sends us back to our hotel for an early night. We leave the next morning for Vienna, Austria, only two and a half hours from Budapest.

CHAPTER 33

Culture Shock

Ricardo's bashed-up Mercedes covered in dust contrasts sharply with the pristine Audis and BMWs gliding through the streets of Vienna. We give a fond farewell to our travel companion, Ricardo Alfredo Silicani from Cordoba Rosario, Argentina, who we have gotten to know well during our week-long road trip. Although we share our home addresses and promise to keep in touch like most of the people we meet on our trip, we will never see or hear from him again. The Argentinean drops us off in the Stephansplatz, the geographical center of Vienna. While removing our packs from the back, Pete calls out, "Wonderful traveling with you, amigo! Muchisimas gracias!"

Stephansdom Cathedral looms over us, its Gothic grandeur a stark reminder of our insignificance in this polished, indifferent city. Although we're excited to explore this fabled city, we feel strangely out of place. Well-dressed people bustle past us. We become self-conscious of our shabby clothes and packs. Vienna is a dressy culture. We see no tennis shoes or hiking boots. No one smiles at us or returns our "guten tags".

Pete and I have always wanted to go to Austria after seeing The Sound of Music, with its cinematic, scenery of magnificent snow-capped mountains. On our three-hour car ride from Budapest to

Vienna the landscape was flat and unremarkable with no mountains. The Alps are near but not visible from Vienna—disappointment number one.

The youth hostel, and the Vienna Opera House are near the center. We head to the hostel to drop off our backpacks.

On arriving, I ask the clerk in English, "May we have a bed for tonight?"

The man behind the desk curtly replies, "Wenn Sie in Osterreich sind, müssen Sie Deutsch sprechen." It is an American assumption that everyone everywhere will speak our language. They don't. Austrians see it as disrespectful that Americans don't even try. I had one year of German in high school and have a preschooler's German vocabulary. I begin again, "Guten tag, Bitte, Koennen wir bitte betten fur nacht haben?"

"Was? Fur heute nacht?'

"Jawohl. Tonight. Heute nacht. Ich spreche nicht gut Deutsch."

"Nein, wir haben keine betten fur heute Nacht."

"Bitte, wo kann man noch?" Please, where is another place we can stay?"

The clerk turns his back on us and says in perfect English, "I can't help you."

In the entire three years of our trip, no hostel refused to help us. In New Zealand, one hostel even set up tents in the back to accommodate the overflow. Some hostels called other hostels or locals to make sure we had a roof over our head. The uncaring attitude of the Austrians—disappointment number two.

In Asia, we were adventurers. In Vienna, we are wanderers without a place--ghosts of a journey now nearing its end. Our new status as lowly road gypsies shocks us.

"It's so strange to come back to the West and become invisible," I say to Pete.

"You're right. To these people we're homeless, poor, and unemployed which is true," he adds.

"It makes me start to worry about getting home and finding a job," I say.

"It makes me feel anxious, too. Traveling has been our job for a year now. I hope we can fit back in the work world. All those strange and different jobs I did in Australia aren't resume builders. I'll have to account for these three lost years some way," Pete worries.

"Hey, we're still on our trip." I remind Pete to lift his spirits, "Let's check out this grand old city of music. We may run into a musical genius."

We decide to walk around the Ringstrasse, the grand boulevard circling the old city. One majestic building after another rise above us showing off a multitude of architecture; Classical, Gothic, Renaissance, and Baroque.

"This was once the center of European music where Beethoven, Mozart, Brahms, Haydn, and Strauss composed and performed their music. I wish I could go back to those times," I say.

"Listen. I hear a piano playing. Let's follow it," Pete replies. We duck down a narrow alley to a coffee shop. There a man is playing Brahms on a grand piano in the storefront window.

"How about this, Rohver? We've gone back in time", Pete says smiling. We buy some coffee and pieces of cake so delicious-looking we can't decide which ones to buy and end up getting more than we can eat. At this moment Vienna is magical.

We walk back toward Stephansplatz. On the way we find Pension Monopol for three hundred and six shillings or $19 a night. Although it's over our budget we rent a room for two nights.

That night, Sunday, we buy student gallery tickets to a Mozart concert at the Vienna Opera House. We climb up five flights of stairs until we're almost touching the ceiling. Standing, we lean against a brass railing for the concert. I have my best outfit on: my mini skirt with a matching top and a pair of hiking boots, which earns looks of disdain from the uniformed staff. The sound is magnificent, and we revel in music we haven't heard for months. This evening the program is Mozart's Symphony No 41, Jupiter, envelopes us in its grandeur.

The next morning, we get up too early to enjoy the pension's continental breakfast, because if you get to Schoenbrunn Castle early enough, the ticket office isn't open and we can sneak in to explore the grounds. We take a fifteen-minute cab ride to the open palace gates. At the entrance we look in awe at this immense, 1,444-room Palace.

"Why does this palace look so familiar?" I ask.

"It was in the movie, The Great Race with Jack Lennon, Natalie Wood, and Tony Curtis that came out in 1965," Pete answers. "I think it's been a backdrop in a lot of other movies and TV shows as well."

"Wow. Interesting," I say, as I look at the magnificent building

Behind the palace is the Great Parterre or sculpted French garden. The garden ends at the colossal Neptune Fountain. In the center of the fountain is Neptune with his trident looking down into a large stone pool. On either side his Tritons (mermen) and raging horses look on. Thetis, a sea deity, kneels at Neptune's feet, but he seems to be indifferent to her pleading.

"This fountain certainly captures the Austrian culture; indifference to lesser beings," Pete notes.

"Do you think Maria-Theresa the empress of the Holy Roman Empire was trying to impress anyone with this place?" I ask.

"Maybe, just a little," Pete says, looking around.

On the hill behind the fountain is the sixty-meter-high Gloriette, an expanse of columns made from recycled war debris. Ironically, it's called Just War, a war that is waged to bring peace. Does adding the word "just" make any war acceptable? I feel it's a rationalization for senseless killing.

Although this was the summer residence of the Habsburg Rulers, it was their preferred residence year-round. Franz Joseph, the longest-reigning emperor of Austria, was born at the palace, liked Shoenbrunn better than his other residences, and died there.

It's snowing. I raise my hands above my head catching the soft glitter falling from the sky like stars filling the night. The prettiness of the snow envelopes us from the cold as we walk back to our hotel. Aren't there moments that are better and sweeter than others?

The next morning, we lie in the soft, heavenly bed until late. The clean sheets, comfortable mattress and warm duvet are luxuries which are hard for us to leave. We shoulder our packs, check out, and head to the dining room. After a breakfast of croissants and coffee we go out into the cold. The light snow of last night has turned into five inches of slushy snow. Ignoring the snow, we tramp off to hitchhike to Salzburg, Austria. Peter can see that I'm not too happy about the situation, so he says, "Kris, Salzburg is the setting for The Sound of Music. Aren't you excited about seeing it?"

"Well, maybe," I answer.

It is wet and cold and cars whizz by, callously passing us. An hour passes. I begin to cry. Teardrops freeze on my face. I relive all the unfriendliness we've experienced in Vienna. What were we thinking expecting a ride from the Viennese when they wouldn't even talk to us? I stamp my feet angrily.

Austria welcomed the Nazis in World War II. These same Nazis sentenced Pete's dad to hang for treason because he was in the Danish underground. In spite of all the sweet cakes, music and lights, there's a heaviness about this city that weighs on us.

"We just need one person to give us a ride. Don't worry. Someone will stop," Pete reassures me. Just then, a man in a Mercedes stops to pick us up and drives us to Salzburg. "Bitte, bitte," I'm afraid we'll melt all over your seats, I say.

"Don't worry, it won't hurt the leather." I give him a hug with my eyes. He speaks excellent English and for the three-hundred mile trip we enjoy good conversation and the toasty warmth in the car.

When we arrive in Salzburg, we find a second-hand store and buy jackets, hats, and gloves. Fall in Europe is cold in the mountains. Hitchhiking isn't going to work so we switch to train travel. We go to the train station and take the first train out of Salzburg to Bern, Switzerland. It's a nine-hour trip. We get on the train at midnight and arrive in Bern at 11 AM. The comfortable seats and gentle rocking of the train put us to sleep.

Once in Bern, we call our Swiss friends, Gisela and Roger Minder, whom we met in Malaysia several months ago. They drive ten miles from

Culture Shock

their small village, Dotzigen, outside Bern to pick us up. It's so welcoming to see friendly and familiar faces after the disagreeable people in Vienna.

We sleep in their loft and spend the nights sharing travel stories and slides. The day after we returned from the Malaysia trip to Singapore, they flew to Los Angeles where they stayed with Pete's family in Calabasas, a suburb of Los Angeles.

The pictures of their trip to Disneyland and the beach with the Fischers make us homesick. I see pictures of Pete's younger sisters, Karin and Mary, for the first time. There are seven Fischer kids; Pete, Anne, Doug, Barbara, Paul, Karin, and Mary. I've been practicing names and ages from Pete's mom's letters that always give family news chronologically. By the time we get to California in December, they will have moved to their new home outside of San Francisco in Diablo. Pete will never return to the rambling ranch house in Calabasas he called home.

During my three years globetrotting, my sister Joan will move from Roanoke, Texas, to Aptos, California, to Omaha, Nebraska, and have a new daughter, Heather. My middle sister will move from Milwaukee, Wisconsin to the Chicago suburb of Hinsdale, Illinois. Only my divorced parents are living in the same place. Baltimore for my mom and Belmont, New York, for my dad. You never go home again after traveling.

The next day, we take the commuter train into Bern. Gisela and Roger tell us to make sure to go to the center of Old Town and watch the changing of the hour at Bern's famous Clock Tower, Zytglogge. It is as old as the city dating from the 13th century. Initially, it was a guard tower at the western entrance to the city. It became a women's prison, a lookout, and a fire observation tower. The city grew around it and since it ended up in the center of the city, they made it into a decorative clock tower. Since 1530 the clock tower has been the official timekeeper of Bern. It chimes out the hour with a precision that only the Swiss could master. Inside the clock tower you see the antique wooden and metal clockworks.

As nine o'clock approaches, we stand beneath the clock tower waiting. At four minutes to the hour a mechanical crowing cock announces the hour is approaching, then several wooden dancing bears frolic on a ledge to the left of the rooster followed by a jester above the bears who jokingly signals the hour too early. All gold Chronos, the God of time, turns his hourglass over and then strikes the hours with his hammer.

"This is quite the show," I exclaim.

"I like the Swiss sense of fun. Let's come back for a second show," Pete says.

Across from the clock tower is the bear pit where live bears have been kept in the city to view and in earlier times for bear baiting contests. Bear baiting is an animal blood sport where a bear is chained by its leg or neck and then packs of dogs are unleashed to attack it. It was popular from the 13th to the 17th centuries in Europe. in 1974 the bears were moved elsewhere so the bear enclosure is empty.

"I like bears and was looking forward to seeing them," Pete says

"Just look around. Likenesses of bears are everywhere; on the clock tower, on the buildings, bridges, flags, and lampposts," I answer.

"Live bears are better," he responds.

The Old Town of Bern has kept its medieval charm and we wander down narrow streets lined by cafes, book, antique, and liquor stores. "I wonder if we can find some wine from 1948 when my parents were married," Pete wonders. We missed their twenty-fifth wedding anniversary and want to buy them something they might like. We find a narrow wine shop with wine bottles on every wall.

Culture Shock

"May I help you?" a gray-haired man with a bow tie and dressed in tweed asks.

"We're looking for a twenty-fifth anniversary present for my parents. We thought an old bottle of wine from 1948 would be perfect," Pete replies.

"That will be difficult. It was right after the war and the wineries in Europe were not producing-much," he tells us. This comment makes us realize what post-war Europe must have been like—no wine, vineyards destroyed and lives disrupted.

"Maybe we can find something down here." He leads us into the cellar and down a tunnel going underneath the city street. He dusts off several bottles, none of which are from 1948, but we see a bottle from 1927 which we can afford. It's old, but Steen and Mary Anne might have drunk it before the war.

On our last night we treat the Minders to dinner at their favorite Gasthof in Hotel Kruez. Their warm Swiss hospitality stands out in stark contrast to Austria's coldness. "We feel we know you better having stepped into your everyday lives for a while," I say.

"We enjoyed staying with Pete's family in the states. You never really know a country unless you get away from the tourist places and stay in a real home," Roger responds. "What do you think about Switzerland?"

"It's clean, picturesque, and has friendly, open people like you two," Pete answers.

The next morning, we leave on the train for Amsterdam to visit the Bergsma Family. I lived with them during the summer of 1966 as an exchange student from New York.

CHAPTER 34

Meeting up with Old Friends in Holland & Scotland

In 1966 I won an AFS (American Field Service) scholarship to the Netherlands for three months. I hit the lottery with my host family, the Bergsmas. They lived in Amstelveen near Schipol Airport, just a bike ride from Amsterdam. Returning to the Bergsma's was like stepping back into a cherished chapter of my youth. Their hospitality reminded me of the summer that anchored me during later turbulent times.

As I write this chapter, I call Jetske, my AFS sister, and pick her brain for details of that visit with Pete in 1974, but we laugh at how little we remember. Once in Europe, I stop journaling, so I pull at my memory to reconstruct the last months of our trip.

It's October 8th already, and 1974 is coming to an end. Time is running out on our year-long trip, so we spend only a few nights with my AFS "father and mother." They give us a belated wedding reception with friends, neighbors, and family. Like all Dutch gatherings, its full of laughter, handshaking, and Jenever (strong Dutch gin). My AFS mother is no longer a housewife, and is working as a law professor while her husband continues to work for KLM.

Many of the guests are from the neighborhood. Every day a neighbor would come to their garden around eleven o'clock for the morning break with strong coffee, sweet pastries, and conversation. It's been eight years since I've seen them, but they still remember much about that summer and our walks and coffees together.

Jetske married Henk Pos, who she was dating when I spent the summer with the Bergsma's. We take the train to visit the couple in Hengelo, where Henk is beginning his medical practice. They live in a huge old house where the medical office is in the front and the living area fills the rest of the house including a walled garden. We'll return and stay in this lovely house ten years later on a European vacation to Holland with our kids and their kids of the same age.

We go for a walk in Beleefbos Park, a wooded area outside Hengelo with no other hikers. Under the autumn sky, we reminisce about old times, each breath visible in the cold air. It feels like no time has passed, yet everything has changed.

"Remember that boating party in Aalsmeer? We rode motorbikes there and almost upset the rowboats with our splash battle?" Jetske remembers. I smile and think back to my summer dating John Lioni, Henk's artist friend. Speeding down the autobahn, my arms tightly wrapped around John made me feel so cool and grown up dating a handsome twenty-six-year-old.

"As I remember it, we drank far too much wine for the ride home, but luckily, we made it home without crashing the bikes," Henk adds.

"How are the Lioni's now?" I ask.

"Like us, he's married now and living happily in Amsterdam. His sister Helen is a wild, and crazy single woman," Jetske answers. That splendid summer in Holland began my travel addiction. The adrenaline of being in new places, seeing new sights, and meeting different people was implanted with this first trip outside Hornell.

We spend only two days in Hengelo, but our connection with Henk and Jetske will last for years. They will visit us in Maine, Montana, Washington DC, and Ecuador, and we will visit them on all our trips to Europe.

Back in Amsterdam we walk around the city visiting the Rijks and Van Gogh museums. The luminous darkness of Rembrandt's Night Watch transfixes as we stand in front of the massive painting.

"When I was here as a teenager, Jetske and I swooped through the museum looking at Rembrandt's most famous painting for a minute and then going to Amsterdam's Red Light district which we were more curious about than old art," I tell Pete.

"Naughty, Kris," Pete says.

A few blocks away from the old masters in the Rijksmuseum is the then-new Van Gogh Museum which had just opened a few months earlier. It has the most extensive collection of Van Gogh's work worldwide. I have just read Lust for Life by Irving Stone. I see the tortured artist in every brush stroke. I buy a print Almond Blossom by the artist. It's one of his few lighter pieces that celebrates the birth of Van Gogh's nephew, Vincent Willem Van Gogh, who would later build the new museum dedicated to his uncle.

We're reluctant to leave the Bergsma's, who have been so kind and caring, but we take the train to Brussels and then the Night Ferry to Victoria Station in London the next day.

Our friend, Robert Thomas, who lived with us in Bundeena, Australia, quits his current farm job, buys a car, and plans a fall trip through Scotland with us. Robert is a compact, rugged guy. His dress is always a camo jacket, shorts, and hiking boots. His refined London accent startles those meeting him for the first time—refined English from this rough-looking guy.

We take the train to Birmingham at Euston Station. He picks us up at the Birmingham Station in his new car, and we head to his family home, Sling Cottage in Romsley, Halesowen. It's the middle of October, and the fall colors are showing. We stay in the guest Trolley House for a night. Robert and his dad have been bachelors for several years since Robert's mom died, and it shows. The house is untidy, but the grounds are lovely, and Robert's mom's garden is still intact.

The next morning, we speed past Birmingham, then Manchester, and stop in Carlisle to see Hadrian's Wall. The Wall is the border between Scotland and England, running seventy-three miles across the island from the Atlantic to the North Sea. It looks different at each location, from turrets to castles to rubble. Britannia fell when the Romans left to defend Rome. How strange it seems to walk this ancient relic of the Romans. Cows and sheep graze around us. Stepping where others stood centuries ago is eerie. This is our Western heritage seeping up through this ground to claim who we are today.

Finally, at Wallsend, Hadrian's Wall ends. We travel toward the Irish Sea coast and hike around Dunur Castle. After a bracing walk, we settle into a small pub in Dunur for some fish and chips and a pint or two. Having lived with Robert in Bundeena, Australia, for several months, I'd learned that his idea of a good meal is a piece of bread soaked in grease, so I don't expect good food, but the fish and chips we order aren't too bad.

The old bar is in a stone cottage with a low ceiling of heavy wood beams charred by years of wood smoke. A pungent smell of sour beer, and loud laughing mixed with live fiddle music fills the room.

Pete, tall and skinny, strides up to the bar with Robert to taste the local brew. Robert whispers to Pete, "Let me handle this, mate. Jimmy, What 'av you on tap? Can we try a nip?"

"Tennants and McEwans," answers the barkeep. "Here's a taste fur ya." Robert and Pete down two shot glasses of beer.

"We'll take the Tennants, Jimmy. Tah. Give us a joug (Scottish pint or three imperial pints)."

Pete adds, "Make mine cold."

"Just stepped off the bloody plane, Yank? No cold beer in this pub.'

"OK," Pete spits out.

"So yous wandered around the castle, did ya? Kennedys owned that castle yonder. Once they barbecued an abbot to get his land. I hear your American Kennedy, Joseph, also knew how to steal a bob or two."

Robert slams a Scottish ten-pound note on the bar, raises his glass, says, "Air do Shlainte!" (To your health), and then walks back to the table.

"Hey, Robert, how did you know the barkeep? You called him Jimmy, and how did he know I'm American?" Pete wonders aloud.

"All Scottish barkeeps are called 'Jimmy.' Did you expect to pass as a local in that get-up? A pair of blue jeans and a worn-out leather cowboy coat? You can't hide that twang either, friend. Americans are all about cold drinks as well. You're as Yank as they come, mate."

A band is playing Scottish folk music. There's a tenor banjo player, a guitar player, and a red-haired angel singing the tale of a country boy who comes to the city and gets himself in a bit of pickle when he falls for a young barmaid and drinks too much. We belt out the chorus "too rah yay" with the rest of the bar. With the dartboard, folk music, and beer, this is a Scottish pub at its best.

The following day, we drive through Glasglow, Scotland's most populous city, and head into Edinburgh, the capital of Scotland, keeper of Scotland's history, and home of Edinburgh University, one of the best universities in the world. The castle overlooks the medieval Old Town and the elegant Georgian New Town. The day is overcast, and the castle looks forbidding and threatening. We walk around it and then retire to another pub. The menu offers haggis, tatties, and neeps. We order tatties and neeps. Neeps are mustard-yellow rutabagas, and tatties are healthy and tasty turnips. When we learn that haggis is a sheep's stomach filled with sheep pluck (sheep liver, lungs, heart, and liver) mixed with suet, oatmeal, and seasonings, we give it a pass. Even Robert of the iron stomach shies away from the haggis.

The next day we're in the Northern Highlands and hiking, hiking, hiking. The highland cattle graze placidly, their shaggy coats catching the amber light of the setting sun. Scotland's wild beauty felt like a fitting prelude to the end of our journey.

Britain has over thirty thousand lakes, and we drive along one of Scotland's deepest and most infamous, Loch Ness. Above its shores are the ruins of Urquhart Castle. It's overcast, and we stop and scan the loch for the Loch Ness monster, a legendary dinosaur-like animal that inhabits the lake. But unfortunately, all we see is Nessie's image in every pub and hostel along the road.

We drive several long hours, and then Robert leaves us off at Birmingham Station where we take the train and ferry to Brussels. We assure our other London friends we'll be back in the near future, but for now we need to get home. It's already the middle of November and we're yearning to be with our families again. For the next twenty years our lives get tangled in careers, mortgages, and children and visits to see London friends never happen.

On an overcast, early morning we leave Brussels on an eight-hour plane flight to New York City. In flight we get reeducated into all the American culture we've missed. We plug into the airline's monitor watching movies we've never seen and listening to musicians we've never heard of. Who's Jim Croce? Check out this movie, Lilies of the Field. After three years of wandering, we're time travelers out of touch with our times.

CHAPTER 35

Homecoming

As the plane touches down, I realize that while our journey has ended, the way we see the world—and ourselves—has changed forever.

"Welcome home," the JFK airport immigration officer says as he stamps our passports. He looks up and smiles, "You kids have been gone a while!"

"Yes," I answer as tears roll down my cheeks. What am I feeling, relief, happiness, sadness? Here we are home--a place we know, love, and understand.

As well-trained travelers having been through Bangkok, Djakarta, and Delhi, navigating New York City is a breeze because we easily read all the signs and understand Americanese. From train to subway we arrive at Penn Station and board the bus to Baltimore.

It's late when the taxi pulls up to 112 St. Dunstan's Road, the castle-like Howell home where my mom has been caring for her aging father. She opens the heavy mahogany door, silhouetted for a moment by the foyer's light, and then races down the walk wrapping her arms around me and my pack. She turns to Pete and makes a debutante-perfect introduction.

The next night nestled in Granddaddy's cozy study sipping his legendary whiskey sours we pull out our presents for them. I wrap the Kashmiri shawl around my mom catching a whiff of her Chanel #5 and give her a hug long enough for my three years gone. Her eyes glisten; a bridge between where I'd been and the home I left behind.

She fingers the delicate pink and purple lotus blossoms edging the wrap and says, "I'm glad you're home, and we're so happy to meet your handsome young man."

I kiss her cheek and make a promise to call her often. Three months prior her mom died and she looks tired, lonely and sad. We leave, become independent, but are always drawn back to our parent's care--that secure nest of unconditional love not knowing they might need a nest as well.

For the next few weeks, we hopscotch across the US, catching up with family and friends; in Belmont, NY, Chicago, and Omaha. My Dad is so excited when he meets us, he backs his new Camaro into a post; talking nonstop about his travels in the Navy. We have sent him a box of our slides from Nepal and we watch them together with his adopted daughter and his new wife. He checked out the slides when he got them in the mail. When we look at them together, he gives humorous guesses as to what and where they were taken, and then we correct him. Over smokes and fishing, Pete and my dad get to know and like each other.

In Chicago, I stay with my sister, Kathy, and then travel to Omaha, where my sisters and Pete and I celebrate Thanksgiving together with their families. Sitting around the large table holding hands as we say grace, I realize how much I've missed my family and their children. Home wasn't just a place we returned to; it was a concept we carried with us; one we would build together in the years to come. Seeing my sisters' families, homes, and lives. I project the future. What now? I hold my older sisters tightly. When my mom fell ill, they were there taking care of their nine-year-old sister. They are giving me advice and a preview of houses, kids, and marriages.

Somehow on our tight travel budget, we managed to buy presents for all our family and friends. Carefully cataloged in the back of my diary are the gifts, their cost, where we found them, and who they are for; Mr. F, ear cleaner (India) $1, pipes and pipe-stand (Istanbul) $14.50: Doug, tooth (Burma) $1, sweater (Kashmir) $8… Although we were gone, our families were never far from our hearts.

Pete's mom picks us up at the San Francisco bus station on the first anniversary of our wedding, December 15, 1974, marking exactly a year after our trip started. We spend Christmas with the Fischer family. I finally meet all six of Pete's brothers and sisters. It's a happy reunion. We stay with the Fischer's until Pete lands a job as a driller trainee on an inland oilfield barge in Louisiana.

Both of our parents talk to us about finally getting the travel bug out of our systems, settling down, working on careers, and having kids. But, little do they know, we we're already looking for our next adventure.

As I look back, our trip seems magical. Twenty-two countries are stamped in our passports. Names pop out of my diary like extras on a movie set; Laurie, Andy, Yam Gurka, Kerry, Pila, Christian, Joe, Yetze, Ben, a watchful girl, an elderly American woman, a Portuguese student, Joe Claven from Temple Road, Mt. Airy Philadelphia, David Moltz, a Peace Corps friend, and his wife to be, a sixteen-year-old groover named Beam J. Carey Jackson from Fountain Valley, California, and Cosmos Blank from Iowa City, Iowa.

My address book is also full of names of people we met and planned to visit sometime in the future which never happened. There was Bill Corkhill from Gundagai, Australia, Andrew Patterson from County Cork, Ireland; Rosie Lloyd from Dorset, England; Laetitia Nemo from Paris, France; Rolf Quaas from Vila de Coude, Portugal, Yani Valencic from Ljubljana, Yugoslavia, the Oakmans from Burnie, Tasmania, and many more.

The intensity and transience of these friendships amazes me. It opened me to new ideas, different perspectives and a love of humanity. I developed a deep tolerance and understanding of people from different walks of life, different ethnicities, and different perspectives on life. Those people and my memories of the times we spent with them and all we gave and shared with each other will stay etched in my heart forever. Most of them we never saw again, but as I read through their names, memories flood back. It's a moving mural of faces, voices, and ideas. In hostels and restaurants, we shared travel advice, our lives, and secrets with an openness I will never experience again.

Pete and I survived and grew in our relationship that year, weathering travel hardships and constant togetherness. Later when we were having trouble in our marriage, we'd remember this trip, one of the best times of our lives, and carry on. December 2023 is our fiftieth wedding anniversary. It's been a great run, and we're still going.

We would leap at the opportunity whenever a chance presented itself to move to a new part of the US or go overseas. Pete never worried about me resisting a move because I'm a fellow nomad, and my bag would be packed before his. As a result, we have lived and worked all over the country and world, from Mohali, India, to Houma, Louisiana. We now divide our time between Cuenca,

Ecuador, Chebeague Island Maine, Augusta, Montana, and Twin Falls, Idaho.

I listened to a podcast interview with <u>Lonely Planet</u> founders Tony and Maureen Wheeler while writing this book. As I listened to their travel story, I felt a strong kinship. They took off from London to Sydney overland a year before we left, following the "hippie trail" through Asia. Writing about their travels in 1973 spurred a multi-million-dollar travel guide business, Lonely Planet. They successfully made travel their livelihood. When asked about his happiest moment, Tony replied, "The first year Maureen and I met, then that big overland trip. That was amazing. I can still remember so much of it so clearly." His words echoed how we felt about our trip.

Adjusting to everyday life in the US wasn't easy for us. We were overwhelmed by the United States' abundance and fast pace. We had no money, credit, or careers, but we had plenty of energy and problem-solving expertise from traveling. Although backpacking around the world isn't an experience you put on a resume, our travels gave us tolerance for different people and ideas. We also developed a persistence in getting things done. As a result, we now appreciate all the opportunities open to us in the US

After three years abroad, we caught up with our peers. We had two wonderful kids, bought a house and more houses, and built our careers. I earned two post-graduate degrees and became a reading specialist in Virginia. Pete finished his career as Chief of Staff for a US Senator. We also have five grandchildren split between New York City and Twin Falls, Idaho.

Fate and serendipity marked much of our trip. Our meeting and then our meeting again was pure luck. Fate led us to some places and not to others. Yet, we look back at the trip and wonder, what if? What if we'd canceled our cruise and gone upriver with the New Guinea trader? What if we'd flown to France and traveled with the erotic artist? What if we'd taken the ship to Greece instead of the car trip to Vienna?

"Pete, What's the best thing you got from our trip?"

"I realized that people everywhere are unfailingly kind when they don't need to be. They were normal people working their way through life, but they took the time to befriend us. What was your big takeaway from our trip, Kris?"

"I learned there's a lot to see and learn from our world through travel," I answer.

The world's circumference at the equator is 24,855 miles. We went 37,400 miles by land, sea, and air. We traveled that distance plus half a globe more. It took us a year to the day, and we spent $5,000 for the entire trip, a bargain. We were poor in 1974, but it was one of the richest years of our lives.

As global citizens, we now see everything differently. Viewing the world from a realistic perspective is humbling and enlightening. We have much to learn beyond our borders. Our trip around the world is over, but traveling will never be out of our systems.

Epilogue

Sometimes large undertakings have small beginnings. This is how I came to write *Travels with Pete*.

On a cool spring day in May Pete and I arrive back in the US from Ecuador to Twin Falls, Idaho. It's 2020 and the world-wide Pandemic is raging. Three days prior we took a State Department emergency evacuation flight home to the US. All borders are closed and our plans to travel in Europe are canceled. We're glad to be back near our son, but are new to Twin Falls. We know no one, not that anyone is socializing anyway. It's a good time to stop procrastinating and unpack all the boxes from our last address in Washington, DC.

As we dig through the boxes, memories come to life in old pictures and saved mementos. I come across a notebook the size of a deck of cards with gold lettering, Diary 1974. Holding the burgundy diary in my hands, I feel a rush of gratitude for the younger self who dared to venture into the unknown. She taught me lessons I continue to carry today.

Pete and I have traveled the world together for fifty-one years. Today we split out time between homes in Ecuador and the United States. It's been a wonderful life full of adventures. We may be slowing down, however, my memories remind me that travel—whether of the body or the soul—never truly stops.

About the Author

After thirty-six years of teaching in Australia, India, and five different US states, Kris Fischer is retired and writing a piece of her own story in *Travels with Pete*. She splits her time between Cuenca, Ecuador; Chebeague Island, Maine; Augusta, Montana; and Twin Falls, Idaho. She's still married to Pete with two children and five grandchildren.

Acknowledgements

My thanks and appreciation to my friends and family who critiqued, encouraged, and supported me in the writing of this book. Through the Cuenca Writer's Collective, led by Franny Hogg, I began writing again. John Keeble and she went through the book's first messy draft and made sense of it. Carolyn Hamilton and Sandra Beaumont taught me about dialogue and adding emotion to my writing. Jeremiah Reardon and Ann Fourt edited my chapter on Vienna online from Cuenca to Twin Falls, Idaho. My four Hornell High School friends; Shirley, Julie, Ed, and Joe, read and encouraged me throughout. My editor, Lynne Klippel, read my mind, adding depth and finishing touches to the book.

I can't leave out my family, daughter, Katy, son, Paul, my sister, Joan, and Pete's brother Doug. They kept pestering me about the book until, after three years, I finished it.

There are many more people who I haven't named here who made constructive comments and helped with computer problems. You are all appreciated.

Lastly, thanks to Pete who corrected and shared memories of our trip and patiently put his life on hold as I wrote.

For photos of this trip, go to my web page designed and managed by Jonathan Mogrovejo, at krisfischer.com.

Made in the USA
Middletown, DE
16 May 2025